ADVANCE PRAISE

"Jaye's work lays bare the depths to which industries must descend before they can self-correct and realize true potential. At a time when we are witnessing unprecedented struggles in valuing nature and respecting Indigenous communities on the front lines, it serves as a stark reminder that the ultimate price for inaction will be ours to bear. This book is both a warning and a call to embrace nature-based economies as the foundation of a sustainable future."

Damien Mander
Founder and CEO, Akashinga

Published by
LID Publishing
An imprint of LID Business Media Ltd.
LABS House, 15-19 Bloomsbury Way,
London, WC1A 2TH, UK

info@lidpublishing.com
www.lidpublishing.com

A member of:

BPR ⚙

businesspublishersroundtable.com

© Jaye Connolly, 2025
© LID Business Media Limited, 2025

Printed by Short Run Press Limited
ISBN: 978-1-917391-33-7
ISBN: 978-1-917391-34-4 (ebook)

Cover and page design: Caroline Li

JAYE CONNOLLY

CARBON CREDIT$ ARE CRAP

THE MYTHS, THE MESS, AND THE WAY FORWARD
FOR TRACEABLE CARBON CREDITS

MADRID | MEXICO CITY | LONDON
BUENOS AIRES | BOGOTA | SHANGHAI

CONTENTS

FOREWORD

Jaye Connolly's *Carbon Credits Are Crap* tackles a difficult topic: many may never have the nerve and courage to pull through the multi-level mess that has been created by trillion-dollar businesses that does not satisfy their greed for more. She exposes its flaws and the injustices it perpetuates against vulnerable communities, particularly in our continent called Africa. However, it is not just related to the African continent, all the exploited Indigenous rural populations are facing similar challenges that call for awakening the world. The worrying factor is the way the carbon credits are packaged and marketed as the solution for climate change; unfortunately it serves as a vehicle for fraud.

The Lambas of Zambia and the Democratic Republic of the Congo, we are among the numerous extensively looted tribes in Africa because of the mineral wealth found on our tribal lands. It is disheartening to watch wealth that is rightfully ours being taken away while we remain in extreme poverty, without any intentions to bring targetable development for us. Then, we have added insults that come in the form of carbon credit funds, that raise the hope of the true stewards of the land, who are left without the financial benefits promised by carbon markets. It is easier to have these experiences as nightmares, and not a constant reality of suffering throughout our lives without hope of reparations or restitutions. The reason is that some of the people involved have built castles on our wealth and without the dehumanizing strategies they will have nothing, and they fear being in our position as their rightful place because the reality is clear that they have nothing without our wealth!

Connolly's argument is compelling, as is her call for transparency, technology-driven accountability and a fundamental rethinking of carbon credits. She proposes cutting out middlemen and ensuring that grassroots communities have direct control over funds that are meant for them. Her solutions include the use of blockchain technology and local government structures to prevent the misappropriation of climate funds.

The greatest value in this book lies in its insistence that true climate justice must empower local communities, not exploit them. Should compensation be given where it is promised, hope will be restored in Lambaland. If the world is serious about fighting climate change while respecting Indigenous rights, then the solutions must be rooted in local leadership, financial transparency and sustainable grassroots economies that resist global predatory practices. Connolly's book is an essential read for policy makers, activists and anyone invested in ethical environmental solutions.

Dr Rosemary Chimbala (PhD) is a distinguished leader in HIV & AIDS research, serving as the Director of the HIV & AIDS Research Unit at Nelson Mandela University in South Africa. Her research interests encompass HIV & AIDS, Indigenous knowledge systems, leadership, inclusive education and education management. She is proficient in multiple languages, including English, French, Cilamba, Kiswahili, Cibemba and Nyanja. She also sees firsthand how the climate of corruption is affecting Africans, particularly in areas related to public health, education and development.

PREFACE

The world stands at a crossroads. In this book, I do not argue about whether climate change is real. Who cares? For far too long, carbon credits have been touted as the magic bullet in this fight, a concept that promised to balance economic growth with environmental stewardship. But, as we have seen over the past 30 years, this promise was hollow, leaving many – especially those in underdeveloped and vulnerable communities – behind.

This book is a critique of the past and a call to action for the future. My journey into the world of carbon credits began with frustration and disillusionment. Since then, it has evolved into something more profound: a mission to expose the truth, uplift marginalized voices and rebuild a broken system. It is about cutting through the noise of greenwashing, holding those in power accountable, and creating a path prioritizing people, the planet and equity.

As you read through these chapters, know my proposed path is difficult. It challenges the status quo, demands uncomfortable conversations and pushes for real change in an industry that has long avoided it. I hope you find inspiration and the courage to act on the message within these pages.

We are all stewards of this planet and are global citizens. It's time to step up. Let us journey to Africa.

INTRODUCTION

TEDX Talk – "Carbon Credits Are Crap"

In 2024, I was asked to give a TEDx Talk titled "Carbon Credits Are Crap," which focused on the rampant fraud in the carbon credit system.[1] It seemed like a straightforward topic to me – after all, carbon credits have been a hot topic in climate circles for years. However, during the talk, I quickly realized that many people had no idea what a carbon credit was. In true TEDx fashion, I could not dive into specifics and had to leave much unsaid. So, here we are – this book delves into the details I could not share on stage.

Let me be clear: I do not care whether climate change is real. What matters is the $2 trillion climate economy. My priority, and the focus of this book, is ensuring that underdeveloped communities – especially those on the front lines of protecting their environment – benefit. Too often, the ones doing the most to safeguard our planet are left with the least.

The company of which I am chairman and CEO – RippleNami – has worked intensely with African nations for a decade, deploying technology to grow their economies. Learning that the most fraudulent practices around carbon credits were happening in Africa hit me hard. This was when my professional focus became a personal mission. I have witnessed fraud entrenched in the carbon credit market and its devastating effects on these communities. While foreign corporations make trillions from carbon credit deals, the communities that should benefit remain impoverished and overlooked.

These pages will explain how the carbon credit system works, why it has failed and what can be done to turn things around. Whether you are familiar with carbon credits or hearing about them for the first time, we will dig deep into what's happening behind the scenes of the climate economy.

CHAPTER 1

THE
PROMISE
AND THE
PERIL

CARBON CREDIT TUTORIAL

Imagine this: for every plume of fossil fuel smoke that goes up into the air, a tree springs up somewhere in Africa. Seriously? For the past 30 years, this has been the bold promise of carbon credits. But what if I told you carbon credits are crap? They look good on paper, but if you dig deeper, you will find unbelievable carbon credit fraud, known as greenwashing.

Now, before you start panicking about checking your carbon footprint app, let me take you on a journey – a 15,000-kilometre journey, no less – to the heart of Africa, where I went searching for answers. I found a complex web of hope and disillusionment, and a level of greenwashing that would make even the savviest marketer blush.

Let's backtrack a bit. Carbon credits are often seen as a silver bullet for the climate crisis. The idea is simple: if a company contributes to global warming, it can offset that impact by investing in projects that reduce or absorb carbon dioxide (CO_2) elsewhere. It's like driving a gas-guzzling SUV but feeling guilt-free because you planted a few trees.

While it may seem like balancing the scales, the reality is far more complex.

Carbon credits function like IOUs in the fight against climate change. They allow companies to continue emitting CO_2 while buying credits to compensate for those emissions elsewhere. It is like balancing a night of indulgence with a healthy salad the next day. Theoretically, it's perfect: for every metric tonne of CO_2 emitted, a credit is purchased, representing a metric tonne of CO_2 removed or prevented from entering the atmosphere.

The idea behind carbon credits is appealing. Companies and individuals in developed countries can carry on as they are without guilt, believing they are offsetting their emissions by investing in projects in places like Africa. These projects either remove CO_2 (e.g. large-scale reforestation projects) or reduce emissions through renewable energy (e.g. solar farms). Each metric tonne of CO_2 offset generates a carbon credit, which can be bought and sold just like a share in a company.

A booming global marketplace approaching $2 trillion[2] actively trades these carbon credits. It is a trillion with a 'T,' as this might be the most lucrative environmental scheme since recycling became mainstream. And let's be honest, the concept sounds pretty great – until you start asking questions like "Where exactly is this solar panel that is supposed to be generating energy in my name?" and "Why does it feel like someone's getting rich, but it's not the local communities?"

LET'S TALK AFRICA

With its vast forests and untapped natural resources, Africa was quickly identified as the ideal location for these carbon credit projects. What better way to help the continent than by turning its environmental assets into a global commodity? But there is a catch – actually, there are several.

Africa

First, let's talk about the scale of Africa. We are not dealing with a manageable plot of land; we are talking about a continent that spans about 30.37 million square kilometres.[3] If you grew up in Texas like I did, you might not be too familiar with kilometres – so, to put it simply, 30.37 million square kilometres is about 44 times the size of Texas or roughly three times the size of the United States. It is an enormous area, about four-fifths of the moon's surface!

Within this vast expanse, you have everything from arid deserts to lush rainforests, and from bustling metropolises to remote villages where technology is more a dream than a reality. Try managing a complex carbon offset project across diverse and often challenging terrains. Spoiler: it is not easy, and things can go wrong faster than a viral tweet in a PR crisis.

Then there is the issue of oversight – or the lack thereof. In theory, these carbon credit projects are rigorously monitored and verified by various governing bodies to ensure they do what they are supposed to. In practice, however, this oversight often feels as rigorous as letting teenagers self-monitor their screen time. There are loopholes large enough to drive an electric truck through and, as a result, many of these projects end up being less effective than the glossy brochures suggest.

And let's not forget the local communities. They are supposed to be the primary beneficiaries of these carbon credit schemes. They are the ones living in and around these forests, and they are the ones who should be seeing improvements in their quality of life due to all this money pouring in. However, instead of new schools, clean water or job opportunities, many communities feel left out of the global climate change conversation. The promised benefits often remain just that – promises.

So, here we are: Africa, a continent rich in natural resources and potential, is stuck in the middle of a global tug-of-war over carbon credits. On the one hand, the developed world is eager to buy carbon credits and claim it is saving the planet. On the other hand, local communities are trying to figure out how they ended up with all the responsibility but none of the rewards. It is a situation as complex as international trade agreements, so I decided to dive into the details and see what is happening.

THE INFAMOUS ARTICLE

But before we get into the nitty-gritty of my investigation, let's rewind a bit. My journey into the tangled world of carbon credits started with a single, eye-opening article. It was one of those pieces that makes you pause and think, "Am I reading this right?"

27 January 2023

Showcase project by the world's biggest carbon trader actually resulted in more carbon emissions

by Bart Crezee and Ties Gijzel
Follow the Money

The Follow the Money headline

On 27 January 2023, my colleague Stanley Mathuram – who is every bit as fascinating as this book – sent me an article titled "Showcase Project by the World's Biggest Carbon Trader Actually Resulted in More Carbon Emissions," written by Bart Crezee and Ties Gijzel of Follow the Money.[4] The article began my deep dive into the murky waters of carbon credits. (Incidentally, don't worry – you'll hear more about Stanley later on in this book. He is a character you will not want to miss.)

The article claimed – brace yourself – that a staggering 94% of forest carbon credits are worthless. Ninety-four percent! For me, this was like realizing that a 'diamond' I had proudly been wearing was cubic zirconia. Men – it does matter. It is shocking and disappointing, and it makes you question everything.

The article detailed how, as part of their well-intentioned (or PR-driven) quests to achieve net-zero emissions, companies had been buying these worthless carbon credits like they were limited-edition sneakers. In one Zimbabwe project, Kariba REDD+, companies unknowingly bought worthless carbon credits to offset their emissions, generating over €100 million in sales for a climate company. The result? About as much real-world impact as a treadmill in a marathon – you are moving but not getting anywhere.

The article also reported that 30%, or over €30 million, of the revenue was said to have gone to the communities working to protect these forests. I initially thought, "OK, bad for the climate and the companies, but good for the communities, right?" But, as I will relate below, this was far from the reality I discovered.

The article got me thinking: What is going on here? How could a system designed to save the planet and uplift the world's most vulnerable people fail so spectacularly at both? 1 am not a climatologist – 1 am the chairman and CEO of a technology company, for goodness' sake – but something about this did not sit right with me. And that is when my professional focus turned into a personal mission.

THE JOURNEY

I embarked on a 15,000-kilometre journey to Zimbabwe to see firsthand what over €30 million in carbon credits looked like on the ground. I wanted to meet the people who were supposed to benefit from these projects and understand how a system that seemed so promising could be so disastrously flawed in practice.

Surviving the Journey to Kariba

Getting to the Kariba area of Zimbabwe is not for the faint of heart, but I made my journey even more complex as I wanted to be able to compare Kariba to other conservation areas. So, I embarked on a 2,600-kilometre road trip with three Zimbabweans and one Kenyan – all initially willing, though I am sure they quickly regretted their decision. We headed south toward the Zimbabwe–Mozambique–South Africa border.

We stumbled upon Gonarezhou National Park, Zimbabwe's wild, remote gem tucked away in the southeast, right on the edge of Mozambique.[5] Gonarezhou means 'Place of Elephants' and this is no exaggeration – it is elephant central! Spanning over 5,000 square kilometres, it is a thrilling playground for big game, such as lions, leopards and buffalo, all set against the jaw-dropping background of the towering Chilojo Cliffs. The park is not your run-of-the-mill tourist destination – it is nature's best-kept secret, offering rugged charm and an authentic wilderness experience with few crowds. Rivers including the Runde and Save slice through the landscape, drawing in an incredible array of wildlife. If you are craving a real adventure and some truly wild encounters, Gonarezhou is the place to be!

The 'causeway' in Gonarezhou National Park

We learned about wild encounters firsthand through a river-crossing story for the ages. Picture this: driving on a 'causeway' (a type of road), crossing a surging river in a non-safari vehicle, with hippos to the right, and crocodiles and elephants to the left – not our most brilliant move, but the stories and videos we got from it were worth every heart-pounding second!

Satisfied with our baseline comparison, we set off north for Kariba – a hidden treasure near the Zambia border, surrounded by rugged terrain that laughs in the face of GPS (and, indeed, there is no internet to enable GPS to even work). Nestled in the northern part of Zimbabwe on the southern shore of Lake Kariba, one of the world's largest artificial lakes, this spot is a haven for those who love nature, conservation and, of course, a good fishing tale. With the lake framed by towering mountains and dense bushland, nature decided to show off a little. Kariba is a hotspot for wildlife, too – think elephants, hippos and crocodiles all spending time together like it is their private resort. And the sunsets? They are the kind that make you wonder if someone secretly cranked up the saturation. You can gaze at the Southern Cross constellation, only visible in the southern hemisphere. Who knew? You do now.

Flat tyre number two

As for getting there, crossing the river was a breeze compared to the 'roads' – if you can even call them that. We managed to rack up a few flat tyres, which the locals fixed faster than a Formula One pit crew, in true safari style. No breakdown call-out service in the bush to rescue us, but who needs them when you have a local with tyre plugs and a winning attitude?

Finally, we reached our ultimate destination: Mola, a remote village in the northwestern part of Zimbabwe's Kariba district, perched on the southern shores of Lake Kariba, within the rugged Zambezi Valley. Mola is a traditional rural community surrounded by challenging terrain and can only be reached by rough, unforgiving roads, thus the flat tyres. Life in Mola revolves around the lake, with the villagers relying on fishing, subsistence farming and the area's natural resources to sustain their way of life. Here, the connection to the land is deep, and the village provides a rare and authentic glimpse into Zimbabwe's traditional lifestyle. Despite the harsh conditions, the people of Mola are renowned for their resilience, warmth and unwavering hospitality.

An outside school room (in Mola), where kids sit in the dirt

We saw the challenging conditions of the local school. What struck me the most throughout this journey was the kindness and happiness of everyone we encountered. And I do mean everyone. People often ask me what I love about Africa. My answer is always the same: it is the people. Despite having next to nothing, they have an incredibly kind spirit. Witnessing such genuine warmth and generosity from those with so little is humbling.

CHIEF MOLA'S SHOCK: CONFRONTING HIDDEN MILLIONS IN CARBON CREDITS

At Mola, which is part of the Kariba REDD+ carbon credit project, we met the chief. Named Champion Rare, Chief Mola is a respected traditional leader of the Mola community. Raised in the area, Chief Mola comes from the Tonga people, an Indigenous group deeply connected to the Zambezi River. The construction of the Kariba Dam in the late 1950s displaced many Tonga families, including those in Mola.

Chief Mola

Chief Mola strongly advocates for his people, focusing on preserving their culture, rights and environment. He plays a crucial role in mediating conflicts, upholding customary law and maintaining social cohesion. Chief Mola's leadership balances traditional ways of life with modern challenges, particularly in environmental conservation and sustainable development. He is the epitome of dignity, standing tall and proud despite his people's daily challenges.

Young girls collecting water, which they do on a daily basis

When I explained the purpose of my visit and shared the article about the Kariba REDD+ carbon credits, Chief Mola's reaction was a mixture of shock, disbelief and frustration. "€30 million, like in US dollars?" he repeated, his voice tinged with disbelief. "If we had €30 million, this place would look like New York City!" He gestured around him, indicating the simple mud homes and the dusty roads. "We need the basics – education, clean water, jobs, food, fuel – and there is no internet around here." The chief shared that, over the past 12 years, his community had received some bags of seed and soil, and the means to construct a sad-looking garden. He explained that his village had dug holes after being promised a borehole water supply, but the necessary equipment had never arrived. There had also been talk of new schools and job opportunities for the local youth, but the promises made had come to nothing. He asked, "Where is this money?"

Chief Mola and his wife, Jessica

It was a good question, but I did not answer it for him then, and I still do not have an answer for him today. The chief went on to tell me how he did not know about carbon credits.

What struck me most about our conversation was the chief's deep sense of betrayal. Here was a man who had trusted the local project developer and foreign climate companies and had his people's best interests at heart. But he had been left with empty words and the faint hope that things would improve someday. I expressed my deep regret to Chief Mola for the climate companies' deceitful actions and assured him that I would take action to help. That promise brings us to where we are today.

As I left the village, I could not shake the feeling that I had just witnessed the tip of the iceberg. If this was happening in one community, how many others were experiencing the same thing? How many people was the system, designed to help, actually leaving behind?

UNSUNG HEROES OF CONSERVATION: EXPOSING THE REAL PROTECTORS OF WILDLIFE

As luck would have it, I accidentally came across Steve Edwards and Damien Mander, individuals who had devoted their lives to protecting Zimbabwe's wildlife and supporting local communities. Both men had spent years on the front lines preserving the environment and ensuring local communities benefitted from conservation efforts.

STEVE EDWARDS:
MUSANGO SAFARI CAMP

With Steve and Wendy Edwards, Musango Safari Camp, Zimbabwe

Steve and Wendy Edwards, the owners of Musango Safari Camp,[6] welcomed me with the warm hospitality you would expect from those who have dedicated their lives to the African wilderness. Nestled on one of Lake Kariba's islands near Matusadona National Park, their camp is a testament to their passion for this incredible landscape.

Pardon the shameless plug for Musango Safari Camp – an unexpected discovery with private chalets that offer a perfect blend of luxury and adventure, each featuring a secluded plunge pool surrounded by the beauty of the African wilderness. After a day of exploration, these tranquil hideaways invite you to relax in comfort while immersed in the sounds of nature – birds twitter in the distance, hippos grunt and splash, and crocodiles thrash in the waters of Lake Kariba, creating an unforgettable atmosphere where the wild comes alive around you. As evening falls, you will join Steve and Wendy for a shared meal under the stars, with the camp transformed into a symphony of the animal kingdom.

Now, back to the story. Wendy is delightful, and it did not take long to decide we must be sisters, bonded by our shared no-nonsense attitude and determination. When I asked how she and Steve had met, Wendy laughed and said, "I met him as he was walking out of jail after killing a poacher – turns out it was someone's son from the previous government administration." Oopsie. Steve then chimed in with a grin, "I'm a tough bird. Once, a hippo grabbed me from my ass and took off into the forest. The locals managed to get me out of the hippo's mouth and locate the only sober pilot within 100 kilometres to life-flight me out."

Lake Kariba tree skeletons, Zimbabwe

Steve is not just a safari camp owner but an enthusiastic conservationist with decades of experience protecting Zimbabwe's wildlife as the former national park head warden. His expertise extends well beyond the local wildlife, encompassing the region's diverse landscapes, ornithology and palaeontology. He recently uncovered a fossil of a new lungfish species, *Ferganoceratodus edwardsi*, in Kariba; it was named in his honour and marks a significant discovery that deepens our grasp of evolutionary history.[7] In the same region, Steve has discovered many other fossils and the only complete dinosaur egg ever found in Zimbabwe, further cementing his role in revealing crucial aspects of the prehistoric life that once thrived there. His wealth of knowledge is based upon the wide range of activities available at Musango, each enriched by his captivating stories and insights. Elon, if you happen to be flipping through these pages, here is a quick heads-up: Steve installed Starlink, and he loves it! In return, he named a hippo after you.

As we sat overlooking the breathtaking landscape, Steve explained the stark realities of wildlife conservation in the region. "What the project developers don't tell you about," he began, "is the real work of protecting wildlife and supporting the community by people like us on the ground, using our funds. These project developers come in, claim to support conservation, and then use our work to sell carbon credits. But the truth is, we do not see a dime of that money as they claim they do not have any money. We did not know we were supposed to receive funding from these projects until you told us. It is a disservice not just to us but to the entire concept of conservation."

Steve's words were a revelation. I had assumed the carbon credit projects were funding the conservation work that was supposedly their backbone, as represented in their eight-by-ten glossy marketing materials. Instead, I learned the developers were riding on the coattails of those genuinely dedicated to the cause, like Steve, without providing the promised support or informing them that they were entitled to it.

DAMIEN MANDER:
AKASHINGA RANGER

Damien Mander (in shorts) and Akashinga

My next meeting was with Damien Mander, a man who needs a little introduction to the world of wildlife conservation. Damien, a former Australian special forces soldier, has dedicated his life to protecting Africa's endangered species. He founded Africa's first plant-based, all-women anti-poaching unit, called Akashinga.[8] Damien is revolutionizing animal protection, community support, and wilderness landscape restoration and safeguarding. He believes change is most effective when it is led by the community, and who better to drive that change than the women who were raised within it? For an inspiring documentary on the Akashinga Rangers, check out *Akashinga: The Brave Ones* by National Geographic.[9]

The Akashinga Rangers

Damien explained that being an Akashinga Ranger means more than wildlife conservation and protection. These women bring income to their communities, creating green economies with positive generational outcomes. They can now purchase property, build homes and send their children to school full time. They also obtain driver's licences, enrol in college and finish their degrees. Akashinga delivers ecological stability and long-term protection of large-scale wilderness landscapes by supporting and empowering local communities.

Damien is a fierce advocate for the role of local communities in conservation efforts. When we met, his passion for his work was immediately evident. "Mate," Damien said, "the biggest threat to wildlife here is not just poaching – it is the lack of real, on-the-ground support. We are out here, day in and day out, using our resources to protect these animals and their communities. But the project developers are nowhere to be seen. They are making claims about their impact on wildlife conservation, but we are doing the work – and we did not even know that we should be getting financial support from these projects. They are taking credit for our efforts without lifting a finger themselves."

Damien explained how his team had funded and executed anti-poaching operations, community education projects and wildlife protection initiatives – often with little to no external support. Carbon credit projects had touted their contributions to these efforts without providing the tangible assistance that was desperately needed. Instead, these projects had taken credit for the hard work and dedication of people like Damien and his team, who risk their lives to be effective while being kept in the dark about the financial benefits they deserve.

Damien added: "I realized finally that conservation is not a conservation issue. It is a social issue – we have a conservation outcome when we have a social impact. That is what makes the value of carbon credits so important. Suppose the money reaches the communities and allows organizations like ourselves to work hand in hand with these efforts. In that case, communities will see the long-term benefits of nature conservation, not just high fences with areas defended by rangers with guns. It will be the people who decide the future of conservation on this continent, which will have two billion people by 2040. Not biceps and bullets."

Hearing this from Steve and Damien was both inspiring and infuriating. Along with their teams, these men are the real protectors of Zimbabwe's wildlife, yet the carbon credit projects were profiting from their efforts without giving back or acknowledging their role. It was clear that something was deeply wrong with the system. The promised funds were not reaching the people and projects that needed them most. Instead, the money was staying in the pockets of project developers and climate companies more interested in selling credits than supporting real conservation work.

These conversations made it abundantly clear that the problem with carbon credits goes far beyond financial mismanagement – it is also a moral failing. The system is built on exploitation, not just of the land and resources but also of the people working tirelessly to protect them. Steve Edwards and Damien Mander are living proof that those on the ground are doing the actual conservation work without the support or recognition they deserve.

I felt a renewed urgency as I left my meetings with Steve and Damien – not just about uncovering the truth behind the carbon credit projects but also about ensuring that the people genuinely making a difference were recognized, supported and empowered. It was time to hold the developers accountable and shine a light on the real heroes of conservation – those who are out in the field every day, doing the work the world desperately needs.

GOVERNMENT OUTRAGE: UNCOVERING THE CARBON CREDIT BETRAYAL

My next stop was to meet with Zimbabwean government officials. Indeed, I thought they would have some answers. I imagined a room full of well-informed bureaucrats, ready to explain the intricacies of carbon credits and how they were being used to benefit the country.

Instead, I found outrage – pure shock, then unfiltered anger. The government officials' expressions hardened when I told them about the article and that a foreign company was making over €100 million, with the communities receiving only €30 million from their trees in the name of carbon credits. It was as if I had told them that someone had stolen their national identity.

"What do you mean they made over €100 million?" one official demanded. "And where is the €30 million promised to our communities? We have not seen a cent of it."

These officials felt as betrayed as Chief Mola and the conservationists. They had not known the air was worth that much money, and now they were left grappling with the reality that they had been exploited. I promised I would assist in educating them and accelerate their participation in the climate economy.

By this point, I felt like I was uncovering a global conspiracy. How could so many people be involved in these projects, with so few benefits making their way to the communities that were supposed to be the primary beneficiaries? It felt like the system was built on sand, and I was determined to uncover the truth. So, I kept on digging.

GATEKEEPERS OF CARBON CREDITS: BREAKING DOWN THE FACADE

My final stop was with the key players in this carbon credit ecosystem – the project developers, the standards governing bodies and the climate investors. Each insisted their projects were unshakeable. They boasted about rigorous standards, meticulous monitoring, and positive environmental and community impact, showing countless glossy photos with ever-changing spreadsheets accounting for the funds.

But as I listened to them, I couldn't help but think, "If the system is so airtight, why are over 90% of these forest carbon credits worthless? And why have the communities seen so little benefit?" It was as if everyone was following the script, but no one was looking at the big picture.

By the end of my investigation, one thing was clear: if carbon credits are supposed to be the future of climate action, then the future looks a lot like the present – full of lofty promises, disappointing results and finger-pointing. The entire system felt like a house of cards, teetering on the edge of collapse, with the people who should be benefiting most left holding the bag.

But don't despair – all hope is not lost! We are just beginning our journey – there is more to uncover, challenge and fix. So, grab your sustainably sourced coffee or a stiff drink, and let's dive deeper into this tangled web of climate economics to see if we can turn those empty promises into real, tangible change.

CHAPTER 2

AFRICA:
A CONTINENT
AT THE
CROSSROADS

Before diving deeper into the complex world of carbon credits, let's start by understanding Africa – a continent as diverse and vibrant as the solutions we are trying to impose upon it. Africa is not just another piece of the global puzzle; it is a sprawling, intricate tapestry of 54 countries, each with its own history, culture and environmental challenges. With over 1.5 billion people, Africa is home to some of the most resilient and innovative communities on the planet. Yet, it is also where the impact of climate change hits hardest, despite the continent contributing the least to the problem.

When you think of Africa, what comes to mind? Vast deserts? Dense rainforests? Bustling cities or remote villages? Images of wildlife, markets or poverty? *The Lion King?* Corruption? Africa is all of these and more. It is a place where ancient traditions meet modern challenges – where the natural world is both a bountiful resource and a constant reminder of what is at stake in the fight against climate change.

It is a land of extremes, from the Sahara Desert in the north to the rainforests of the Congo Basin in the centre; and from the savannas of East Africa to the bustling urban centres of Cairo, Johannesburg and Lagos. Within this vast and varied landscape, the people of Africa are as diverse as the land itself. With over 2,000 languages spoken,[10] countless ethnic groups and a rich tapestry of cultures, Africa is anything but a monolith.

Yet, despite its diversity and richness, Africa faces challenges that are as monumental as its size. The continent is home to some of the world's fastest-growing economies, but it is also where poverty, political instability and underdevelopment are most deeply entrenched. I commend Africa for its remarkable embrace of technology; often I experience Africa outpace the US thanks to the absence of outdated technology systems. Africa's technological achievements have allowed the continent to leapfrog over more developed nations in technological advancement.

Over 60% of the population relies on agriculture for their livelihoods, often through sustainable farming practices passed down through generations. However, the majority of Africa's population lives on less than $2 a day[11] – a stark reminder of the inequalities that persist in the global economy.

A CONTINENT RICH IN RESOURCES, STRAINED BY INEQUALITY

NATURAL RESOURCES

Africa's wealth of natural resources is as immense as its landscapes. The continent holds 30% of the world's known mineral reserves, 8% of the world's natural gas and 12% of the world's oil reserves.[12] Oh, and let's not forget Earth's second lung, the Congo Basin, which, alongside the Amazon, generously provides 20–28% of our oxygen[13] – you know, for free. But the majority of the oxygen we breathe (about 50–80%) comes from marine plants like phytoplankton. Africa's vibrant waters, particularly along the West African and Horn of Africa coasts, account for around 10–15% of global phytoplankton.[14] They are also rich in wildlife and home to species you won't find anywhere else.

Yet, despite all these riches, Africa remains the poorest continent, with many people living in extreme poverty. Makes you wonder, right? Let's dive into why.

COLONIZATION:
BRINGING 'PROGRESS' ONE STOLEN
RESOURCE AT A TIME

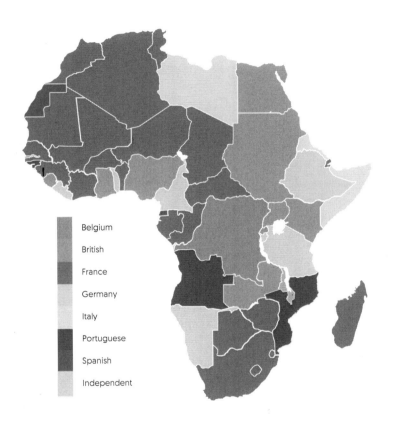

Belgium

British

France

Germany

Italy

Portuguese

Spanish

Independent

Africa's colonization by European empires, 1880–1914

For over 1,300 years, Africa was exploited through the Arab slave trade, which, unsurprisingly, took a toll on the population and resources. Fast-forward to the late 19th century and European powers launched the Scramble for Africa, carving up the continent for their benefit. Belgium, Britain, France, Germany, Italy, Portugal and Spain participated.[15] They were not after African culture or innovation; no, they had their eyes on Africa's wealth of natural resources – gold, diamonds, rubber, ivory and more.

Britain colonized regions across the continent, including Botswana (Bechuanaland), Egypt, Eswatini (Swaziland), Gambia, Ghana (then the Gold Coast), Kenya, Lesotho (Basutoland), Malawi (Nyasaland), Nigeria, Sierra Leone, Somalia (British Somaliland), South Africa, Sudan, Tanzania (Tanganyika, later merged with Zanzibar), Uganda, Zambia (Northern Rhodesia), Zimbabwe (Southern Rhodesia), and parts of Cameroon and Libya after World War II.

France claimed large portions of West and North Africa, including modern-day Algeria, Mali and Senegal, while Portugal laid claim to Angola and Mozambique. Germany colonized Cameroon, Namibia and Tanzania (then part of German East Africa) before losing its colonies after World War I.

Spain held smaller colonies such as Equatorial Guinea and Western Sahara, while Italy took over Eritrea, Libya and Somalia. Belgium took control of the Congo, where brutal exploitation became infamous.

I have purposely saved Belgium for last. One day, I sat with a gentleman for nine hours at an African state house – equivalent of the White House in the US. Both of us were waiting to meet with an African president. As we spoke, I asked him why he was there, and he explained he was running for president of the Democratic Republic of the Congo (DRC) and needed this president's support.

He urged me to read *King Leopold's Ghost*, written by Adam Hochschild.[16] The book truly grasps the dark, brutal history of the DRC – a country that still suffers today despite sitting on an estimated $27 trillion worth of unmined minerals. That conversation has stayed with me as I reflected on the challenges facing Africa and its people. In the end, the gentleman won the election, but the results were rigged and so, despite winning, he 'lost.' It is another tragic chapter in the DRC's long, painful story – perhaps one for another book.

Back to the history lesson. the African European colonization lasted for 70–80 years. European powers justified their actions by claiming they were bringing 'civilization' – a convenient excuse for extracting resources and wealth. Many African nations only gained their independence in the 1950s and 1960s, meaning they have had just 60–70 years of self-governance after centuries of foreign control.

To put things into perspective, 60–70 years after its independence, the US was busy building the Transcontinental Railroad, establishing extensive networks of roads and canals (such as the Erie Canal), and becoming a global powerhouse. After the same amount of elapsed time, whereas the US was chugging along with economic growth and industrialization, many African nations are still picking up the pieces from centuries of exploitation.

And this exploitation seems set to continue. Numerous African countries still deal with the fact that global powers and corporations swoop in, extract their resources and leave without contributing much to local populations.

FOREIGN AID:
THE ULTIMATE CHARITY CON JOB

Meanwhile, foreign aid organizations are notorious for offering temporary fixes instead of long-term solutions. The exploitation of foreign aid in Africa is a harsh reality that often goes unspoken. Let's go there now.

While billions of dollars are raised and lent to help the continent, much of that money *never* reaches the people or projects it's intended for. Organizations like the International Monetary Fund (IMF), the United Nations (UN), the United States Agency for International Development (USAID), the World Bank and others have been criticized for perpetuating a system that benefits lenders and donors far more than the people of Africa.

The IMF and World Bank often tie their loans and aid packages to strict economic reforms – reforms that frequently lead to austerity measures, privatization of resources and the crippling of local economies.[17] These conditions, known as 'structural adjustment programmes,' force African governments to cut essential public services like healthcare and education, all while repaying high-interest loans. The result? African nations remain trapped in cycles of debt, paying back far more than they ever received in aid while basic needs go unmet.

Speaking of the World Bank, it needs an introductory course titled 'Where Is the Money?'! According to an Oxfam audit published in October 2024, a shocking $41 billion, representing 40% of all climate funds disbursed by the bank from 2017 to 2023, has disappeared.[18]

Leading financial contributors to the World Bank, including the US (15.85%), Japan (6.84%), China (4.42%), Germany (4.0%) and the UK (3.75%), collectively invest billions of their hard-earned taxpayers' money each year. In 2023 alone, the bank's initiatives pooled $23.5 billion from over 50 high- and middle-income countries, alongside capital market financing and the bank's funds. But where is the oversight from these nations to ensure these funds are used effectively and transparently? Lapses in stringent oversight have led to significant financial mismanagement and threatened essential development and climate resilience projects globally.[19]

The mismanagement of $41 billion is particularly egregious considering the potential impacts on communities facing the brunt of climate change. These areas have acute needs relating to clean water, healthcare and education. The outrage caused by the Oxfam report stems from the misallocation of funds and the squandered opportunity to significantly uplift and transform lives in the Global South, where communities are most affected by climate change yet least responsible for it. These funds were meant for projects that could have built climate resilience, averted devastating floods or tackled desertification – essential efforts toward sustainable development. The disappearance of these funds is not merely a failure of accountability; it represents a monumental lost opportunity to make a lasting positive impact on some of the world's most vulnerable populations.

But the problem doesn't stop there. The real show happens on the ground, where aid workers, often associated with organizations such as USAID, UNICEF, Oxfam and the UN Development Programme, live a very different reality from those they claim to be helping. I have personally observed aid workers dining at the nicest restaurants in town, sipping wine while their drivers wait in white Toyota Land Cruisers (why are the cars always white?), catching a nap before chauffeuring their exhausted charges back to their very comfortable homes, where an entire staff waits to help them. Oh yes, they are working so hard for 'foreign aid.' Meanwhile, the communities they are 'helping' live without access to power, the internet, clean water, primary healthcare or proper education.

Aid is frequently funnelled through high-priced foreign consultants who charge exorbitant fees, all while knowing little to nothing about the local context. Projects are started but remain unfinished or are poorly managed or unsustainable. Meanwhile, the locals see only a fraction of the intended benefits – half-built schools, crumbling roads and empty promises. To top it off, many African governments have become dependent on these foreign flows rather than building sustainable, self-sufficient systems.

Foreign aid creates dependency and reinforces external control instead of promoting development. And let's not forget that it is not just about charity; assistance comes with a heavy dose of political leverage and with many economic strings. African nations are routinely pressured to align with the geopolitical interests of the donor countries, while local needs are conveniently sidelined.

As a result, Africa continues to suffer from underdeveloped infrastructure, poor healthcare and inadequate education systems. Meanwhile, aid organizations such as the IMF, the UN, USAID and the World Bank paint a rosy picture of progress to justify their actions and fundraising, while on the ground, the reality is far less glamorous. The people of Africa are stuck in a system that benefits the lenders and donors more than it helps the local communities.

It is time to face the uncomfortable truth: foreign aid, in its current form, does more to enrich and empower the 'assisting' organizations and governments than it does to help the intended recipients. Real change requires more than just donations or loans – it demands a fundamental shift in how aid is structured and delivered. The focus must be empowering local communities and governments to drive their development and be free from exploitation and external control. The current model is more about keeping the aid workers comfortable than genuinely helping needy people.

THE GREAT CHINESE 'FRIENDSHIP' IN AFRICA: EXPLOITATION DISGUISED AS AID

China's involvement in Africa is often presented as a partnership – after all, they build roads, railways and infrastructure while offering loans. What's not to love, right? But, looking closer, it is clear this 'friendship' is more about exploitation than solidarity. China has mastered the art of debt-trap diplomacy, resource extraction and gaining political control while keeping a smile on the surface.

DEBT-TRAP DIPLOMACY: LET'S CALL IT WHAT IT IS

China loves lending money to Africa, but this is not because it is generous. The idea is simple: loan countries far more than they can afford, then watch them struggle to repay. When they cannot, China steps in to 'help' – which usually means taking control of strategic assets, ports or other valuable parts of the economy.

Countries like Djibouti, Kenya and Zambia are drowning in Chinese loans. Kenya's new Mombasa–Nairobi Standard Gauge Railway, built with a Chinese loan, has not exactly been the goldmine it was promised, but don't worry – China will collect that debt.[20] You might say China is playing Monopoly – except in this version, they always win, and Africa is left with the bill.

RESOURCE EXTRACTION: AFRICA'S ATM FOR CHINA

At the heart of China's interest in Africa is something simple: resources. Copper, cobalt, rare earth elements and food – Africa has them and China wants them. In true imperial fashion, Chinese companies, many state-owned, secure sweet deals to extract these resources at bargain prices. The profits? They go straight back to China, while African communities are left with environmental disasters and little else.

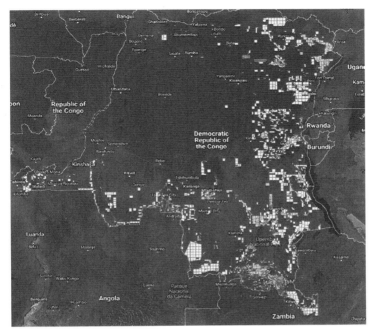

Mining operations in the Democratic Republic of the Congo

Take the DRC, for example. Chinese companies dominate cobalt mining, which is essential for electric car batteries. China gets cheap minerals, but the people of the DRC get dangerous working conditions and environmental devastation. But hey, the West feels good about its 'green' technology.

INFRASTRUCTURE PROJECTS WITH STRINGS ATTACHED: BUILT BY CHINA, FOR CHINA

So, China is building all this infrastructure in Africa – railways, airports and ports. It must be suitable for the locals, right? Not exactly. The catch is that China does not just fund these projects; it also builds them using Chinese workers and materials. So, while Africa gets a new railway, most of the jobs and economic benefits stay with China.

Let's not forget the inflated prices. Many of these infrastructure projects – like Kenya's Mombasa–Nairobi Standard Gauge Railway or Zambia's Kenneth Kaunda International Airport – are massively over budget. When they do not deliver the expected returns, Africa is left with the debt while China watches the interest pile up. It is like lending Africa a ladder but charging them for each rung. Eventually, China will own the assets.[21]

POLITICAL INFLUENCE: AFRICA, SAY HELLO TO YOUR NEW BEST FRIEND

In addition to profiting from Africa's resources and debt, China is building political influence across the continent. It wants not just Africa's minerals but also its votes. China gains political leverage by cosying up to African leaders with loans and investments, often pushing these nations to support Chinese interests in global forums such as the UN. So, next time you see an African country siding with China on the international stage, remember it is not about ideology but the price tag.

CRIMINALS RUNNING THE SHOW

It doesn't end there. China has even sent criminals from its prisons to Africa to manage projects and other Chinese interests.[22] Of course, China refutes the claims, but its on-the-ground workers talk openly. Yes, that's right. Instead of being rehabilitated, they are sent to oversee infrastructure projects, mining operations and business ventures across Africa. Imagine convicted criminals being put in charge of billion-dollar projects and resources. For China, it is a win–win – low-cost labour and a quick way to keep its interests in check.

ENVIRONMENTAL AND SOCIAL IMPACT: A MESS FOR AFRICA, PROFITS FOR CHINA

China's exploitation is not just about money and votes; it is also about land. Its mining and infrastructure projects are tied to massive environmental damage. Forests are destroyed, rivers polluted and farmland ruined – all in the name of progress, which just so happens to primarily benefit China.

And what about the workers? Chinese companies have been accused of paying locals next to nothing while imposing poor working conditions. Meanwhile, Chinese labourers are brought in to do the high-paying jobs. African workers protest, strike and demand better treatment, but the Chinese executives are living comfortably in their air-conditioned offices.[23]

IS CHINA A FRIEND OR JUST ANOTHER MODERN-DAY COLONIAL POWER?

China's relationship with Africa is like that too-good-to-be-true guy on a dating app. He looks impressive online – hot with those abs. But when you meet in person, you realize you've been scammed – those photos must be from 30 years ago. Sure, Africa gets new infrastructure, but it is handing over resources, land and political influence in return. Worse still, it is left with crushing debt that future generations will be paying off long after China has moved on to its next 'investment.'

If Africa wants to break free from this cycle, it will take more than just saying goodbye to outside powers. Africa needs to demand fairer deals, protect its resources and ensure that development benefits go to its people. Right now, China is smiling while Africa is stuck paying the bill. This extractive economic model has deprived Africa of the wealth generated by its resources and left behind environmental wreckage and social dislocation. Poverty amid plenty – that's the irony.

THE CURRENT CARBON CREDIT CIRCUS IN AFRICA

But wait, there's more. In theory at least, the need to shift to a more sustainable, equitable model has been recognized for 30 years. Enter carbon credits. We've already touched on how these work and explored some of their issues, and the next chapter will focus on their mechanics in detail. But let's briefly take a deeper look at how they relate to Africa.

The concept is simple: companies in developed countries pay to offset their pollution by investing in projects that supposedly reduce or remove carbon from the atmosphere in developing nations, such as those in Africa. These projects, from reforestation to renewable energy, sound great in theory. The result? Developed countries get to pat themselves on the back, while developing countries are meant to benefit from investment. A win–win, right? Well, only if the investment makes a difference.

In 2024, in the voluntary carbon credit market, 133 out of 195 countries (68%) participated in 8,000 carbon credit projects, generating 1.8 billion metric tonnes of carbon credits, most of which were worthless, as we'll explore. Notably, three countries – China, India and the US – account for 51% of all projects and carbon credits as shown below.

Carbon credits issued by country

Source: Ivy S. So, Barbara K. Haya, Micah Elias. (2023, May). Voluntary Registry Offsets Database v8, Berkeley Carbon Trading Project, University of California, Berkeley.

In Africa, 40 out of 54 countries (74%) participate in 1,600 projects, generating 230 million metric tonnes of carbon credits. As the second largest continent in the world, Africa contributes only 13% of the total carbon credits and 20% of the projects.[24]

BREAKING DOWN AFRICA'S CARBON CREDIT PROJECTS

The carbon credit landscape in Africa is characterized by a diverse array of projects, each with its own goals, challenges and impacts. The figure below provides a full breakdown.

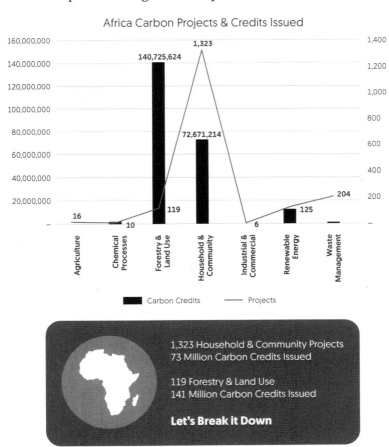

Summary of Africa's carbon credit projects[25]

Source: Ivy S. So, Barbara K. Haya, Micah Elias. (2023, May). Voluntary Registry Offsets Database v8, Berkeley Carbon Trading Project, University of California, Berkeley.

HOWEVER, THREE TYPES OF PROJECT DOMINATE THE SCENE: FORESTRY AND LAND USE, HOUSEHOLD AND COMMUNITY, AND RENEWABLE ENERGY.

Category: Forest & Land Use	Category: Household & Community	Category: Renewable Energy
Type: REDD+ 125 Million Credits	**Type:** Cookstoves – 54 Million Credits	**Type:** Hydropower – 5 Million Credits Wind power – 4 Million Credits Solar – 3 Million Credits
3% of the projects generating	43% of the projects generating	
55% of the carbon credits	23% of the carbon credits	

Combined – 12 Million Credits

Top three types of 2024 carbon credit project in Africa
(see *Chapter 4* for more information on REDD+)

FORESTRY AND LAND USE

Forestry and land use projects are the most significant, with 125 million carbon credits issued for forestry, of which over 90% have been deemed worthless.[26] By my calculations, these projects represent only 3% of the total projects in Africa but account for a staggering 55% of all carbon credits issued on the continent. They focus on preserving existing forests and reforesting degraded lands, crucial for carbon sequestration and biodiversity conservation. The forests in Africa, particularly in the Congo Basin, play a vital role in absorbing CO_2 and are a critical component of global climate regulation. However, these projects are not without challenges. Ensuring the permanence of these forests (explored further below) is a significant concern. Additionally, there are often disputes over land ownership and the distribution of benefits, which can lead to conflicts with local communities.

HOUSEHOLD AND COMMUNITY
Household and community projects are the most common type of carbon credit project in Africa, accounting for 43% of the total. These projects have issued 54 million worthless carbon credits for cookstoves, representing 23% of the total credits on the continent. These projects typically involve distributing energy-efficient cookstoves, reducing the need for firewood, and decreasing deforestation and carbon emissions. They also have significant health benefits for communities by reducing indoor air pollution, Africa's leading cause of respiratory illness. However, the success of these projects depends on their adoption by local communities, which can influence costs, cultural preferences and the availability of alternatives.

RENEWABLE ENERGY
Renewable energy projects are another important category, though they represent a smaller share of the overall carbon credits issued. These projects have issued 5 million carbon credits from hydropower, 4 million from wind power and 3 million from solar power. Africa has vast

potential for renewable energy, particularly solar and wind power, due to its abundant sunshine and wind resources. These projects aim to reduce the continent's reliance on fossil fuels and provide clean energy to local populations. However, they face challenges related to infrastructure, financing, and integrating renewable energy into existing power grids.

THE DOUBLE-EDGED SWORD OF CARBON CREDITS

While carbon credits are often paraded around as a solution to environmental and social good, let's call them what they are – crap! The financial gains from these projects typically line the pockets of foreign investors, project developers and intermediaries. At the same time, the local communities, who are supposed to benefit the most, are left with little to nothing. These communities endure the highest environmental and social costs without seeing a fair share of the benefits. It is exploitation, plain and simple, dressed up in the language of sustainability.

Another challenge is ensuring the 'additionality' of carbon credit projects (the idea that they are genuinely newly created, explored further in the next chapter). This is particularly difficult to prove in regions where development is rapidly changing. Moreover, projects may have multiple benefits, making it hard to attribute carbon savings to a carbon credit investment alone.

Furthermore, the issue of permanence is a significant concern on the African context. How do you ensure that a reforestation project or a forest conservation initiative will continue to deliver carbon savings 50 or 100 years from now? What happens if there is a drought, a fire or a change in land use? And who is responsible for ensuring the project is maintained and that the carbon credits issued are legitimate? These are questions that the carbon credit market has been grappling with since its inception, and they are particularly relevant on a continent like Africa, where natural and human-made challenges abound.

AFRICA'S ROLE IN THE GLOBAL CARBON MARKET: A COMPLEX REALITY

On the one hand, Africa's vast natural resources make it an ideal location for carbon offsetting projects. Its forests, wetlands and grasslands have the potential to sequester significant amounts of carbon, and its abundant sunshine and wind resources offer enormous potential for renewable energy projects. On the other hand, Africa's governance, infrastructure and equity challenges make it difficult to ensure that these projects deliver on their promises.

One of the biggest challenges is governance. In many African countries, the institutions responsible for overseeing land use and natural resource management are weak, under-resourced and often plagued by corruption. Ensuring carbon credit projects are implemented transparently and equitably is challenging. In some cases, projects have been approved without the consent of local communities, leading to conflicts over land and resources.

Another challenge is the lack of infrastructure. Many carbon credit projects, particularly those in remote areas, require significant infrastructure investments to be successful. Infrastructure includes everything from bridges to roads to internet connectivity and power lines. In many cases, these investments are beyond the reach of local communities and require external financing, which can lead to dependence on foreign investors and project developers.

Equity is another primary concern. In theory, carbon credit projects are supposed to benefit the local communities – they are the ones who live on the land. But in practice, the situation is often far more complex. Land ownership in Africa can be murky, with traditional land rights frequently conflicting with modern legal systems. Add in the interests of foreign investors, governments and non-governmental organizations, and you have a recipe for confusion, conflict and exploitation.

Carbon credit projects make grand promises, such as new schools, better healthcare, clean water and job opportunities for local communities. But, as I discovered on my journey (see *Chapter 1*), fulfilled promises are often as elusive as the rain in the Sahara.

As I delved deeper into the world of carbon credits, I increasingly began to identify the complexity of Africa's role in the global carbon market. The biggest challenge of all is perception. Many in the developed world see Africa as a blank canvas – a place where environmental solutions can be implemented without the complications of Western bureaucracy. However, Africa is not a blank canvas. It is a complex, living, breathing continent with history, challenges and aspirations.

As I have learned, imposing solutions from the outside without fully understanding the context is a recipe for failure. In the next chapter, we will explore the mechanics of carbon credits – the good, the bad and the downright baffling. We will explore how these credits are created, traded and verified. And we will closely examine some key players in this global market. So, buckle up because the ride is about to get even more uncomfortable.

CHAPTER 3

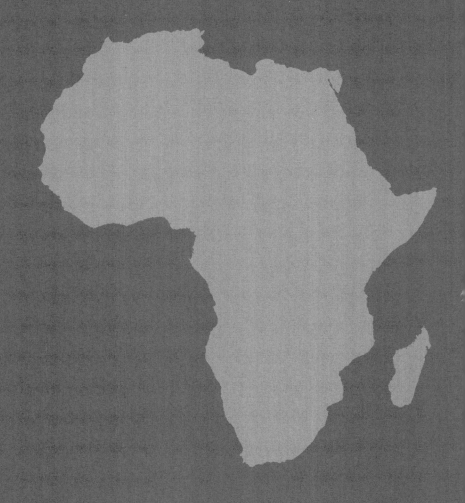

THE MECHANICS OF CARBON CREDITS

You might be wondering, how exactly do these carbon credits work, and why should I care? These are fair questions, especially since the concept of carbon credits can feel as opaque as a legal contract written in fine print. But don't worry – I am going to break it down for you in a way that's memorable and easy to understand.

HOW CARBON
CREDITS WORK

Let's start with the basics. Carbon credits permit the holder to emit CO_2 or other greenhouse gases. One credit typically equals one metric tonne of CO_2. Companies, governments and individuals who need to emit more CO_2 than a specified amount can buy carbon credits from those who have reduced their emissions. It is like ordering a double bacon cheeseburger with a diet soda – you are trying to balance the guilt, but deep down you know you are not fixing the problem.

The idea is that by creating a market for carbon credits, we can incentivize governments and companies to reduce their emissions. If you can make money by emitting less CO_2, why wouldn't you? It is a climate economy where environmental sustainability intersects with economic growth – think capitalism with a conscience, or so the theory goes.

But how do these credits come into existence? That is where the process becomes more intricate. Projects that reduce, remove or avoid emissions of greenhouse gases generate carbon credits. These projects vary widely, including planting trees, capturing methane from land-fill and investing in renewable energy. Each project must undergo a supposedly rigorous assessment to verify that it reduces emissions before carbon credits are issued.

Once a project is up and running, it generates carbon credits based on the amount of CO_2 it is expected to reduce or absorb. Project developers sell certified carbon credits on the carbon market, where companies, governments and individuals can buy them to offset their emissions. Theoretically, this is a win–win: the project gets funding, the environment and communities benefit, the seller makes money from reducing emissions and the buyer feels good about reducing their carbon footprint.

But here's where things start to get tricky. The carbon market is not a single, centralized system. Instead, it is a patchwork of different markets, each with its own rules, standards and players. Most carbon credits are traded within compliance markets, which surpassed $850 billion in value in 2023.[27] Compliance markets, such as the European Union Emissions Trading System (EU ETS), dominate this space by enforcing stringent regulations that mandate emissions reductions across key industries like energy and manufacturing. In the regulated compliance markets, carbon credits are typically issued by governments or international bodies (such as the EU ETS), and the rules are strict. Regulators set a cap on the amount of CO_2 companies can emit, requiring them to buy credits if they exceed that limit. Over time, authorities lower the cap, pushing companies to reduce their emissions or pay higher credit prices.[28]

Initially smaller than the compliance markets, the voluntary carbon market is skyrocketing. Analysts project its value will reach about $100 billion by 2030 as more companies actively choose to offset their carbon emissions to meet net-zero climate commitments. By 2050, the voluntary market could surge to around $250 billion, reflecting the intensifying global focus on combating climate change.[29] The voluntary market is a bit like the Wild West. As we say in Texas, there are a bunch of big hats with no cattle. Here, the rules are looser and the players are more varied. Companies buy credits to offset emissions voluntarily, driven not by legal requirements but by motivations such as corporate social responsibility, public relations or the desire to appeal to environmentally conscious consumers.[30]

However, whether we are discussing regulated compliance or voluntary markets, trust is the key to a successful carbon credit system. Buyers need to trust that the credits they are purchasing represent real, verifiable reductions in emissions. And that's where things can go wrong.

WHERE THINGS GO WRONG

Not all carbon credits are created equal. Some projects are rigorously monitored, with detailed reports and third-party verification to ensure real emissions reductions. But for others, well, the oversight can be a bit lax. Fraud cases in the carbon credit market are widespread (as we will see in the next chapter). Issuers have distributed credits for projects that failed to deliver promised emissions reductions or counted the same reduction multiple times, undermining trust and the credibility of carbon offset programmes. Stricter oversight and verification standards are crucial to restoring integrity to the market. For companies that are the victims of such fraud, it is a bit like buying a luxury Rolex watch only to find out later that it is a cheap knockoff called a Rolix.

Then there is the issue of 'additionality,' a fancy term that means the emissions reductions from a project should be 'extra' and would not have happened without the carbon credit investment. In other words, companies should not receive carbon credits for funding the protection of forests that were already being preserved. Awarding credits in these cases floods the market with ineffective offsets and compromises the real impact of carbon reduction initiatives. Such practices skew the purpose of carbon markets and reward actions that offer no additional environmental benefits, weakening the system's credibility. Ensuring additionality is crucial for the integrity of the carbon market, but proving it can be difficult, especially in regions where development is rapidly changing.

Let's also not forget the importance of longevity of emissions reductions. If a company funds a reforestation project and earns carbon credits for the CO_2 the trees will absorb, what happens if those trees are cut down a few years later? Who is responsible for ensuring that the reductions are permanent? Projects set aside some of their credits as a 'buffer' to account for potential losses, but this is far from a perfect solution.

As I delved deeper into the mechanics of carbon credits, I began to see that the system, while well intentioned, was fraught with challenges. From issues of verification and additionality to concerns about permanence and equity, the carbon market is anything but straightforward. It is a complex web of players all trying to tackle one of the most pressing issues of our time.

But the most surprising thing I discovered was the sheer scale of the carbon market. What started as a niche idea has ballooned into a trillion-dollar industry, with everyone from multinational corporations to small non-governmental organizations (NGOs) vying for a piece of the pie. And, while there is no doubt that the carbon market has the potential to drive real change, the question remains: is it working? The answer is *no!*

THE PLAYERS AND THEIR INTERESTS

Let's now examine the players in this global market – the project developers, the governing bodies, the investors and the communities. We will explore their roles and motivations, and the impact they are having on the ground. Because if we are going to understand the promise and the peril of carbon credits, we need to understand the people behind them.

PROJECT DEVELOPERS

Think of project developers as the architects of the carbon credit world. They identify potential projects, design them and then get them up and running. These projects can be as diverse as reforesting a degraded landscape, installing solar panels in a rural community or capturing methane from a landfill. Once the project is operational, the developer is responsible for monitoring and verifying the emissions reductions, which are converted into carbon credits.

Project developers can be large multinational companies or small NGOs. Their motivations range from pure profit (carbon credits can be sold on the market for a tidy sum) to a genuine desire to combat climate change and support sustainable development. However, the balance between profit and purpose leads to conflicts of interest, particularly when the pressures of making a project financially viable clash with the need for rigorous environmental standards.

In my travels, 1 met with several project developers. Some were enthusiastic environmentalists who saw carbon credits as a way to fund the projects they had always dreamed of. Others were more pragmatic, viewing carbon credits as one of many tools in the broader effort to address climate change. And then some were in it for the money, plain and simple. They were not shy about admitting that their primary goal was to maximize profits, and if that meant cutting a few corners, so be it.

This diversity of motivations is part of what makes the carbon credit market so complex. On the one hand, you have developers who are genuinely committed to making a difference; on the other, you have those who see carbon credits as just another commodity to trade. The challenge is ensuring that the market incentivizes the correct behaviour – projects delivering real, lasting benefits for both the environment and the communities involved.

STANDARDS GOVERNING BODIES

If project developers are the architects, then the standards governing bodies are the building inspectors. These organizations are responsible for setting the rules and ensuring that carbon credit projects meet specific standards. They issue the certifications that allow projects to generate carbon credits, and they play a crucial role in maintaining the integrity of the carbon market.

Several major standards governing bodies exist, each with their own rules and methodologies. The most well-known include the Clean Development Mechanism (established under the Kyoto Protocol),[31] the Gold Standard[32] and the Verified Carbon Standard.[33] These organizations develop the methodologies projects must follow to generate carbon credits and provide third-party verification to ensure the credits are legitimate.

But here is the thing: while these organizations play a critical role in maintaining the integrity of the carbon market, they are fallible. As I explored the world of carbon credits, I found that the rigour and transparency of these standards can vary. Some organizations are known for their strict adherence to lofty standards, while others have been accused of being more lenient, especially when there's money on the line.[34]

One of the biggest challenges for standards governing bodies is verification. Ensuring that a project genuinely delivers the emissions reductions it claims can be difficult, particularly in remote or conflict-prone regions. In some cases, the data used to verify emissions reductions is outdated or incomplete, leading to credits being issued for reductions that never occurred.

Then, there is the issue of conflicts of interest. In some cases, the organizations responsible for verifying projects are also involved in their development or financing. They may even receive payments for every carbon credit they issue. Talk about a conflict of interest! This creates a situation where the fox is guarding the henhouse, raising significant questions about the entire system's credibility.

CLIMATE INVESTORS

Next, we have the climate investors – the financiers of the carbon credit world. These companies, governments and individuals buy carbon credits to offset emissions. For many, the motivation is straightforward: they want to reduce their carbon footprint and contribute to the fight against climate change. For others, it is about corporate social responsibility or, in some cases, simply meeting regulatory requirements.

Climate investors come in all shapes and sizes. Some are large multinational corporations with significant emissions to offset. Others are smaller companies or even individuals who want to make a positive impact. In recent years, we have also seen the rise of impact investors, who are explicitly looking to fund projects that deliver financial returns and social or environmental benefits.

But while climate investors' motivations are often noble, the reality of the carbon market can be anything but. As I discovered, not all carbon credits are created equal, and it can be challenging for investors to know whether the credits they buy are delivering the promised benefits.

The lack of transparency and oversight in the market means it is too easy for investors to be misled, mainly when the pressure is on to meet sustainability targets.

Sometimes, investors may not even realize that the credits they are buying are dubious. They may rely on third-party ratings or certifications without digging deeper into the project's specifics. Because the carbon market is so complex, it can be difficult for even the most well-intentioned investors to navigate.

COMMUNITIES

Last but certainly not least are the communities – the people who live on the land where carbon credit projects are developed. In many cases, these communities are supposed to be the primary beneficiaries of carbon credit projects. They stand to gain from improved infrastructure, job creation and environmental protection.

But as I travelled through Africa, I saw firsthand how reality often falls short of promise. In a very few cases, communities are consulted and involved in developing projects, and see tangible benefits. But in the majority, they are left out of the decision-making process altogether. Outside interests suddenly manage the land they have lived on for generations, and the promised benefits fail to materialize.

One of the biggest challenges for communities is the issue of land rights. In many parts of Africa, land ownership is complex and often contentious. Traditional land rights can conflict with modern legal systems, and who can decide how the land is used is not always clear. Land leads to disputes and, in some cases, outright conflict.

Even when communities participate in developing carbon credit projects, the benefits can be unevenly distributed. In some cases, most of the profits go to the project developers, investors and intermediaries, with little left over for those living on the land. This creates a situation where the environmental benefits of a project are achieved at the expense of social equity – a trade-off that many argue is unacceptable.

BIG PICTURE TIME

So, where does this leave us? The carbon credit market is a complex and often contradictory world, where profit motives, environmental goals and social justice intersect in ways that are not always easy to untangle. On the one hand, carbon credits can drive significant investment in projects that can deliver real and lasting benefits for both the environment and the communities involved. Conversely, the lack of market transparency, oversight and equity means those benefits are not always realized.

As I continued my journey, the need for greater accountability in the carbon market became increasingly apparent. Allowing companies to offset their emissions is insufficient; those offsets must deliver the promised benefits. Achieving this requires removing as many consultants as possible and leveraging technology to eliminate the 'human' fraud effect. Implementing rigorous standards, ensuring transparent reporting, and emphasizing social equity and environmental protection are essential for creating a system that genuinely works and withstands scrutiny.

The next chapter will examine some scandals and frauds that have rocked the carbon credit world. From fraudulent credits to projects that have failed to deliver on their promises, we will explore the darker side of the carbon market and what it means for the future of climate action.

CHAPTER 4

THE
WALKING
INTO THE
MIRAGE

Like any system involving large sums of money, good intentions and global impact, the carbon credit market was bound to attract its share of scandals and frauds. And boy, has this been the case. What was once seen as a noble endeavour to save the planet has tarnished its reputation with a series of high-profile cases that have revealed how fragile – and sometimes fraudulent – the system can be.

Before we dive into the juicy details, let us set the stage. The carbon credit market operates on a delicate balance of trust, transparency and good faith. When these elements are compromised, the system wobbles like a tightrope walker – and it can quickly become apparent there is no safety net. Unfortunately, it doesn't take much for that tightrope to start swaying dangerously, and when it does, the consequences can be significant. One wrong step and the entire system can come crashing down, leaving a trail of broken promises and missed opportunities.

2023: ALL THE KING'S MEN COULDN'T PUT HUMPTY DUMPTY BACK TOGETHER AGAIN

In the unfolding saga of environmental conservation, the year 2023 emerges as a watershed moment. It witnessed the unmasking of widespread malpractices that have tainted the integrity of carbon credit schemes. It became apparent that a phenomenon dubbed 'greenwashing' – the art of cloaking corporate misdeeds in a veneer of environmental concern – had become rampant, casting long shadows over genuine efforts toward sustainability.

A series of exposés showed just how far we have to go in our journey toward environmental integrity. Many of these stories challenged our perceptions and forced us to confront the uncomfortable truths of our time, raising an urgent call for ethical vigilance in the ever-evolving story of our planet's future. Investigative journalism by renowned media houses Bloomberg,[35] Follow the Money,[36] *The Guardian*,[37] *The New Yorker*,[38] SourceMaterial[39] and Survival International[40] played a pivotal role in uncovering the underlying realities of these projects. The headlines varied from "Showcase Project by the World's Biggest Carbon Trader

Actually Resulted in More Carbon Emissions" (Follow the Money) to "Blood Carbon: How a Carbon Offset Scheme Makes Millions from Indigenous Land in Northern Kenya" (Survival International) and "The Great Cash-for-Carbon Hustle" (*The New Yorker*).

Key revelations included the fraudulent nature of over 90% of rainforest carbon offsets[41] and the discovery of critical shortcomings in the world's largest soil carbon removal project, leading to its suspension.[42] Additionally, the reporting highlighted the production of millions of fake emissions credits through cookstove offsets, which were inaccurately certified by reputable voluntary carbon credit programmes and sold to top-tier global corporations.[43] These findings cast a shadow over the legitimacy and efficacy of voluntary carbon credit initiatives worldwide.

KARIBA REDD+ FOREST PROJECT: THE FIRST TO FALL

ROAD TRIP REQUIRED

As discussed in *Chapter 1*, on 27 January 2023 I read an article by Follow the Money that examined the Kariba REDD+ forest project in Zimbabwe and its association with the Swiss-based climate consultancy South Pole.[44] The article claimed South Pole had sold over €100 million worth of dubious emissions rights to hundreds of prominent companies, including Ernst & Young, Gucci, McKinsey, Total and Volkswagen. At first glance, I found the revelation shocking. Most of the article highlighted how South Pole allegedly deceived these global corporations. What particularly piqued my interest was the reference to an estimated €30 million benefit designated for the local communities. For remote African communities living in the bush, €30 million would be life changing. Intrigued and sceptical, I embarked on my Zimbabwe journey.

KARIBA PROJECT
OVERVIEW

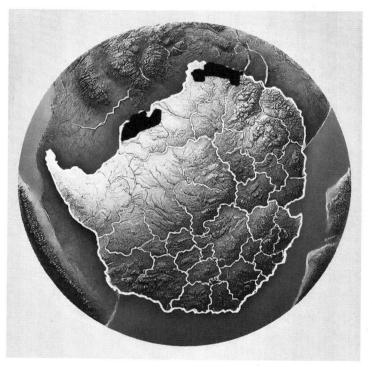

Zimbabwe Map with the Kariba REDD+ project area denoted in black

Kariba REDD+ was a large-scale carbon offset initiative situated along the northern border of Zimbabwe, neighbouring Zambia along Lake Kariba. The project aimed to protect over 750,000 hectares of forested areas from deforestation and degradation.

The Kariba project started in 2011 and abruptly ended in 2023. It aimed to achieve its goals by promoting sustainable land management and conservation practices among the local communities. These practices were designed to prevent deforestation, which helps to reduce carbon emissions. The project generated revenue by selling carbon credits, representing the carbon emissions avoided through these conservation efforts.

Apart from its environmental goals, the Kariba project's goal was to focus extensively on community development and wildlife protection. It aimed to protect wildlife and improve the livelihoods of the local communities through various initiatives such as creating employment opportunities, improving access to education and healthcare, and supporting sustainable agricultural practices.

As a REDD+ project, Kariba was part of a global effort to value and monetize forest carbon stocks, offering a way to incentivize forest conservation and support sustainable development in economically challenged regions. However, like many large-scale carbon offset projects, it faced scrutiny and challenges, particularly regarding the actual effectiveness and fairness of its implementation.

PROJECT MANAGEMENT

The Kariba REDD+ project was managed on the ground by Zimbabwe-based Carbon Green Africa[45] and initially financed by Carbon Green Investments (CGI),[46] a parent company set up in the tax haven of Guernsey. Swiss-based South Pole[47] was in charge of the technical project development work and sale of credits. Verra is a Washington, DC-based non-profit that oversees the voluntary verification of carbon offsets and the issuance of the Kariba project carbon credits.[48] Please remember Verra – they are the certifiers of the majority of crappy carbon credits.

Many consultants were involved in Kariba, but I want to call out a particular group. SCS Global Services,[49] a California-based third-party verifier, played a role in verifying the project's credits. Its involvement in validating a project that ultimately failed to deliver on its promises raises serious questions about the verification processes behind carbon credit certification. These layers of consultants, including South Pole, SCS Global Services and Verra, contributed to a system where communities remain disenfranchised and carbon credits are often fraudulent, benefiting those far removed from the local realities.

According to the project's documentation, the local communities in the Kariba project area, including the Binga, Hurungwe, Mbire and Nyaminyami rural district councils, were the project's owners and primary beneficiaries.

FRAUD EXPOSED

In 2023, South Pole whistleblowers (and later investigative journalists) revealed that over 90% of the carbon credits issued were fraudulent and that this had led to an overstatement of the project's impact on carbon emissions reduction. Additionally, there were concerns about the local communities, wildlife protection and the lack of tangible benefits both had received. Initially hailed as a showcase for environmental conservation and community development, the project faced criticism for failing to deliver its promises and contributing to greenwashing.

This scandal highlighted the challenges and pitfalls in implementing and monitoring large-scale carbon offset projects without complete traceability and transparency. In the crucible of accountability, the Kariba project emerged as a case study, its promise tarnished by broken vows and shattered dreams. Yet, beneath the surface, amid the glitter of the 30 million carbon credits sold to top global companies and a purported €100 million in revenue, lies a stark reality mired in opacity and neglect. A mere fraction of the vast revenue (€30 million out of a total of €100 million) was pledged to uplift the local communities and rural districts within the project's ambit. However, the reality saw this fraction reduce to a pittance. During my visit to the local communities, it became evident that the amount of life-changing funds distributed was significantly less than initially claimed, amounting to under €1 million.

Moreover, the project overestimated its climate benefits while delivering less financial support than promised to the Zimbabwean communities and poaching the wildlife. The project, which had been a symbol of progress in carbon emissions reduction among its corporate clients, is now seen as a risk to the credibility of the entire carbon market. The collapse of Kariba REDD+ impacted the market's insurance mechanisms, such as the credit buffer pool, which is essential for guaranteeing climate benefits in the face of unforeseen setbacks such as wildfires and droughts.

OVER A YEAR LATER

Having generated €43 million in revenue, South Pole removed its CEO and ended its partnership with CGI while maintaining that it had not overestimated the carbon credits. After significant scrutiny of the project's business practices, it terminated its relationship with CGI.[50] It was found that most of the project's proceeds intended for communities and wildlife protection had gone to CGI, not the local communities tasked with fighting deforestation and anti-poaching efforts as initially claimed.

According to South Pole, CGI had received €57 million, of which €30 million had gone to the communities. Despite ongoing controversies, CGI maintained its stance of having committed no wrongdoing, while its CEO remains in the UK, far away from Zimbabwe. Having received consulting fees plus 10 cents for each carbon credit certified, totalling over $3 million, Verra removed its CEO and put the project on hold pending further investigation.

The Lake Kariba communities still live without the basics, and the chiefs banded together and sued all the parties involved. Sadly, the termination of the Kariba project will have long-term implications for the carbon market, the economic sustainability of local communities and wildlife protection. While the local management entity, CGI, faltered in prioritizing community welfare, Verra's flawed audits and South Pole's complicity further compounded the injustice. As of early 2025, the communities are experiencing a two-year-long drought, which is leading to poaching to make up for failed crops and no resolution to their legal suit.

SAFEGUARDS

A pivotal question looms large: who safeguards the interests of these marginalized communities? As the spotlight shifts to accountability and traceability, the absence of robust governance structures and transparent mechanisms for benefit distribution casts a long shadow over the project's integrity. Community benefit-sharing, a cornerstone of carbon credit initiatives, demands more than lip service – it necessitates genuine engagement, inclusive decision-making and stringent oversight.

As the layers of deception unravel, the spectre of unfulfilled promises casts a pall over the landscape. Why did the lifeblood of prosperity elude the grasp of those it was meant to empower? In the intricate web of stakeholders – CGI, SCS Global Services, South Pole and Verra – each bears a share of culpability in this saga of broken trust.

In the labyrinth of carbon credit projects, where profits collide with principles, the Kariba REDD+ project stands as a cautionary tale – a stark reminder of the perils that lurk beneath the veneer of environmental altruism. As the echoes of disillusionment reverberate, it falls to us to heed the lessons of the past and chart a course toward a future where integrity reigns supreme. The promise of progress is not just a fleeting mirage in the desert of deceit.

NORTHERN KENYA RANGELANDS CARBON PROJECT: A PARADIGM OF PROMISE AND CHALLENGES

PROJECT OVERVIEW

The Northern Kenya Rangelands Carbon Project (NKCP)[51] is a significant initiative for carbon removal and conservation in northern Kenya. Situated in the country's vast and diverse grasslands, it focuses on enhancing the carbon sequestration capabilities of the soil. The project began in 2009 and aims to capture carbon and improve the grasslands' overall health by encouraging sustainable grazing and land management practices. It involves 14 of 43 community conservancies with over 175,000 people and focuses on soil conservation techniques (such as improving pastoralist grazing practices and reseeding native grasses), generating additional revenue and enhancing conservation efforts. It covers about 1.9 million hectares and aims to remove 50 million tonnes of CO_2 over 30 years. It also aims to create a scalable model that can be replicated in similar ecosystems, potentially impacting global carbon reduction.

A vital aspect of the project is its focus on community involvement. Local communities, often pastoralists who depend on these grasslands for their livelihoods, are engaged in planning and implementation. The project provides them with training and resources to support them in adopting sustainable grazing practices, which in turn help to improve their quality of life by ensuring healthier ecosystems and more robust grazing lands.

PROJECT MANAGEMENT

The primary party involved in the NKCP is the Northern Rangelands Trust (NRT).[52] Ian Craig founded the NRT after transforming his family's 62,000-acre cattle ranch into the Lewa Wildlife Conservancy (which became widely known after Prince William chose it as the location for his proposal to Kate Middleton).

Craig has continued to play a significant role in the NRT, holding the position of Chief of Conservation and Development. The transition from a cattle ranch to a wildlife conservancy underscores the shift in focus toward conservation and sustainable land management in the region. Over 40 organizations fund the NRT. The United States Agency for International Development (USAID) is a significant contributor, having invested close to $32 million in northern and coastal Kenya since 2004.[53] And, you guessed it, Verra has been involved overseeing the voluntary verification and issuance of carbon offsets.[54]

CHALLENGES AND
CONTROVERSIES

Despite its potential, the NKCP has faced several challenges. A Survival International investigative report, *Blood Carbon*, found that third-party validators assigned to evaluate the project had identified over 100 'findings,' which essentially meant 'concerns,' during a review process.[55] Despite these issues, Verra, the certifying body for the project, eventually verified the carbon credits produced by the project.

Another significant issue is the difficulty of accurately measuring and verifying the amount of carbon sequestered in the soil. This challenge raises questions about the efficacy and credibility of the carbon credits generated.

Additionally, there are concerns about the long-term sustainability of the project and its impact on biodiversity. Most importantly, the project's dubious claims regarding its positive impact on local communities have raised questions about the credibility and ethical implications of such carbon offset schemes. This controversy underscores the complex dynamics involved in large-scale conservation and carbon offset projects, particularly when they intersect with the rights and livelihoods of Indigenous peoples.

The Indigenous communities in the area, including the Borana people, have raised significant concerns and criticisms. In 2023, in response to the *Blood Carbon* report, the Borana Council of Elders released a statement.[56] Borana leader Abdullahi Hajj Gonjobe also released a video denouncing the devastating impact of the NRT's conservancies on their pastoralist way of life.[57] The Borana people strongly condemned the project, labelling it a "green scam" and demanding that NRT vacate their community land. The statement alleges "gross human rights violations by NRT against the Indigenous pastoral communities in northern Kenya," including killings and rape; it said the Borana people "categorically state that no free prior and informed consent process was followed and obtained from our communities." The Borana Council is considering "instituting further legal actions against NRT."

ECONOMIC IMPACT AND CARBON TRADING

The NKCP facilitates the generation of carbon credits, which are subsequently sold on international carbon markets. The soil carbon credit mechanism serves as a revenue stream for the project and supposedly for the communities involved in its implementation. However, as we have seen, local communities have reported not receiving the intended benefits. This monetary incentive is crucial for the project's sustainability and is a model for linking environmental conservation with financial benefits.

From the project's inception in 2013 until February 2023, Verra verified and issued over 6.7 million carbon credits, with around 4.5 million being purchased for offset purposes. Notably, significant purchases include 180,000 credits by Netflix and 90,000 by Meta. USAID reported that in February 2022 the project's 14 conservancies received $5.1 million in disbursements (around $345,000 per conservancy). The conservancies dispute receiving any disbursements. The report stated, "Forty percent will enhance sustainability of community conservancy operations and the remaining 60% will meet community needs through the carbon community fund."[58] The conservancies dispute the fact on disbursements.

KENYA: A STEP TOWARD
A GREENER FUTURE

The NKCP, while pioneering in its approach to combating climate change, faces significant challenges. Despite its potential for ecological conservation and economic development, its effectiveness and impact on local communities remain in contention. Addressing these challenges is crucial for realizing the full potential of such environmentally significant endeavours.

UNMASKING THE REALITIES: THE COOKSTOVE CARBON CREDIT CONTROVERSY

COOKSTOVE CARBON CREDITS OVERVIEW

Cookstove carbon credits are part of the carbon offset market. Improved cookstove projects are the fastest-growing carbon credit project type for a significant portion (around 15%) of projects in the voluntary carbon market.[59] Specifically, these credits are associated with the distribution and use of improved cookstoves in developing countries. The idea is to replace traditional cooking methods (often involving open fires or inefficient stoves) with more efficient, less polluting cookstoves. These improved stoves are designed to burn less fuel and produce fewer emissions, reducing greenhouse gases like CO_2.

The primary objective is to reduce carbon emissions from household cooking and improve local air quality, thereby also bringing health benefits to communities that traditionally rely on biomass (e.g. wood or charcoal) for cooking. To generate carbon credits, project developers must follow specific methodologies set by carbon registries such as the Clean Development Mechanism or the Gold Standard. These methodologies define how to estimate the emissions reductions from using improved cookstoves.

The calculation involves estimating the amount of biomass saved by using efficient cookstoves and converting this saving into equivalent CO_2 reductions. This process can be complex and varies depending on the type of biomass, the efficiency of the cookstove and the usage patterns of households.

Project developers actively sell these carbon credits on the voluntary carbon market. Companies and individuals buy them to offset their carbon emissions, contributing to their sustainability goals.

TOP PLAYERS

According to a *Financial Times* analysis, British Airways, easyJet, E.ON and Shell are among the most significant corporate buyers of the popular cookstove credits.[60] In the world of cookstove carbon credits, some developers might have gotten a little too enthusiastic with their numbers. C-Quest Capital and UpEnergy have both been flagged by the carbon credit market watchdogs for overestimating their carbon credits. And yes, again, Verra has been involved in controversies about certifying carbon credits from cookstove projects under the Verified Carbon Standard. It turns out that assuming how much wood people *would have* burned without their shiny new stoves is not as straightforward as they thought. While these projects aim to reduce deforestation and carbon emissions, they have been known to rely on some optimistic maths – leading to a bit of 'carbon credit inflation.' So, if you're cooking up numbers, ensure they're not overdone!

CHALLENGES AND CRITICISM

Cookstove carbon credit projects face criticism for issues related to the accuracy of emissions reduction calculations, the actual adoption and sustained use of the cookstoves, and the impact on deforestation and local air quality. Recent studies, like one from the University of California, Berkeley, suggest that the reduction in emissions and the health benefits may be significantly overestimated.[61] In September 2023 this study was challenged by a group of experts, who wrote an open letter (published by C-Quest Capital) criticizing the study for its "misguided criticism."[62] Nevertheless, the Berkeley exposé highlighted the over-crediting of cookstove carbon credits, indicating that the actual reductions in greenhouse gas emissions might be much lower than claimed. This raised concerns about these credits' effectiveness and integrity in the carbon offset market.

BUSTED

Finally, someone is being held accountable in the world of carbon credits. In October 2024, Kenneth Newcombe, former CEO of C-Quest Capital and a board member of Verra from 2007 to December 2023, faced criminal charges. The US Department of Justice indicted him and Tridip Goswami, former Head of Carbon and Sustainability Accounting at C-Quest, for wire fraud and commodities fraud. Investigators accused Newcombe and Goswami of manipulating project data to exaggerate the performance of their clean cookstove projects, inflating carbon credit outputs and defrauding investors out of over $100 million.[63]

Authorities uncovered the scheme during an extensive audit of carbon credit projects, aided by whistle-blowers from within the carbon market who provided crucial insights. Investigators found that Newcombe and Goswami had falsified data on cookstove usage and inflated emissions reductions, enabling them to issue far more carbon credits than justified. These revelations exposed significant vulnerabilities in carbon credit verification processes, particularly as Verra, the certifying body under the Verified Carbon Standard, played a central role in endorsing these questionable credits.[64]

Newcombe's dual role as CEO of C-Quest Capital and board member of Verra raises severe concerns about conflicts of interest and the transparency of the certification process. Verra, already under scrutiny for other controversies, now faces mounting pressure to strengthen its auditing and project verification protocols. Beyond fraudulent numbers, what remain missing are the promised community benefits – improved air quality, health outcomes and economic support – that these projects are supposed to deliver to local populations.

The indictment of Newcombe and Goswami underscores the need for stricter regulatory measures, transparent data reporting, and independent verification to ensure the integrity of carbon offset programmes. It signals that more consequences may be on the horizon. As of February 2025, the legal proceedings against Newcombe and Goswami are ongoing. I wonder if any recovered monies will reach the communities these projects were intended to help.

GERMANY'S $5 BILLION CARBON CON: CHICKEN FARMS AND FAKE PROJECTS

Let's dive into the $5 billion carbon credit fraud that rocked Germany in 2023. It's a wild ride, with the scandal centring on carbon reduction projects in Azerbaijan, China and Nigeria. These projects were supposed to help big emitters, like Shell, meet their mandatory reduction targets by investing in carbon credits abroad. But things went off the rails when investigations revealed that many of these projects were utterly fabricated or had hugely inflated their emissions reduction claims. For instance, Shell ended up funding what it thought was a legitimate carbon reduction initiative but that turned out to be a chicken farm. Yes, a chicken farm!

The fraud was uncovered in 2023 when German authorities investigated irregularities in the country's carbon trading system.[65] They found a sophisticated scheme involving shell companies, falsified documentation and the resale of non-existent carbon credits. The credits were sold across Europe and beyond, with many buyers unaware they were purchasing worthless credits.

The fallout from this scandal has been severe, with several high-profile arrests and a significant loss of confidence in the European carbon market. The size of the fraud is staggering – 27 projects in China alone were found to have massive irregularities. And it didn't stop there; other projects across Azerbaijan and Nigeria were also part of the scheme. The failure to detect these issues earlier has raised serious questions about the oversight of Germany's carbon credit system, especially since it allows companies to offset up to 1.2% of their emissions through these foreign projects.

What makes this even more troubling is the lack of stringent verification processes. Verra (again) approved many of these projects despite their fraudulent nature. Are you cottoning on to the issue yet? Investigators even used satellite imagery to show that some of the land tied to these carbon credits was either empty or showed no signs of the environmental impact the credits were supposed to represent.

Aside from the financial hit, the reputational damage is enormous. The involved companies are now under scrutiny, and the loss of confidence in carbon credit schemes has spilled out from Europe to the rest of the world. It has become clear that these fraudulent projects could continue to undermine global climate efforts if complete traceability and transparency mechanisms aren't implemented.

The $5 billion fraud has highlighted the system's vulnerabilities and the ease with which they can be exploited by those looking to make a quick profit. It has also raised fundamental questions about the role of regulators and the effectiveness of existing safeguards.

SYSTEMIC FAILURES AND PERVERSE INCENTIVES

As I delved deeper into these scandals, a disturbing pattern emerged. Carbon credit fraud is everywhere, but many fraudulent projects are located in regions with minimal oversight and weak governance. With its vast and often remote landscapes, Africa is a prime target for these frauds. The lack of infrastructure, the complexities of land ownership and the challenges of monitoring large-scale projects have made it easy for unscrupulous developers to exploit the system.

However, the problem is not limited to Africa. The carbon credit system is built on shaky foundations in Asia, Europe, North America and South America. The standards governing bodies, which are supposed to ensure the market's integrity, are often outdated, overly complex and easily manipulated. Consultants like Verra, which thrive on the system's complexity and opacity, perpetuate these issues by focusing on maximizing profits rather than ensuring environmental integrity.

The carbon credit market is a product of the neoliberal economic model that has dominated global policy for the past few decades. The model prioritizes market-based solutions to social and environmental problems and assumes that the private sector, driven by profit, will naturally find the most efficient and effective ways to address issues like climate change.

In theory, this makes sense. By putting a hefty price on carbon, the market incentivizes companies to reduce their pollution emissions and invest in carbon projects that mitigate the impacts of climate change. But in practice, the market has proven to be anything but efficient or effective. The problem is that the incentives created by the carbon credit market often prioritize short-term profits over long-term sustainability.

For example, project developers are often paid based on the number of carbon credits they generate, not on the actual impact of their projects. This creates a strong incentive to maximize the number of credits issued, even if that means cutting corners or exaggerating the benefits. At the same time, the lack of rigorous oversight means it is too easy for developers to game the system, issuing credits for projects that do not deliver real emissions reductions.

THE ROLE OF STANDARDS GOVERNING BODIES AND CONSULTANTS

The standards governing bodies, such as the Gold Standard and the Verified Carbon Standard, were established as non-governmental organizations (NGOs) to ensure that carbon credits are credible and verifiable, and that they contribute to real emissions reductions. However, these NGOs have become increasingly archaic over time, hampered by bureaucratic inertia and a significant lack of innovation.

The problem lies in the fact that NGOs often resist change due to their reliance on established processes and the involvement of multiple stakeholders. Their rigidity stifles innovation, making it difficult for these organizations to adapt to innovative technologies or methodologies that could improve the effectiveness and transparency of the carbon credit market. They continue operating on outdated models that others easily exploit, undermining their original environmental protection mission.

NGOs rely heavily on a sprawling network of consultants, many who are more invested in maintaining their relevance and income streams than driving meaningful change.

The processes for verifying and certifying carbon credits have become convoluted, slow and disconnected from on-the-ground realities. These issues create an environment ripe for fraud and manipulation by those prioritizing profit over genuine environmental impact.

A network of consultants has the carbon credit market in a chokehold, with zero motivation to innovate. These folks keep pushing the same tired playbook and are more interested in gaming the system than fixing it. Their consultant-driven culture keeps the market expensive, neutral, and allergic to change and fresh ideas.

INNOVATION
AND REFORM

So, where do we go from here? The scandals and scams that have rocked the carbon credit world have exposed serious flaws in the system and created an opportunity for reform. As the market matures, there is a growing recognition that greater oversight, transparency and accountability are needed to ensure that carbon credits deliver the real and lasting benefits they offer.

One of the most promising developments in this regard is the rise of technology-driven solutions that can improve the monitoring and verification of carbon credit projects. Real-time satellite imagery, advanced data analytics and AI-driven monitoring systems are some tools that have been explored to create a more transparent and reliable carbon market. These technologies have the potential to provide real-time data on the impact of carbon credit projects, making it harder for fraudsters to game the system and easier for buyers to trust that their investments are making a difference.

Survival International's *Blood Carbon* report, along-side the Berkeley investigation into cookstove credits, underscores the critical need for these advancements. These reports have shown that the current system is vulnerable to exploitation, and urgent reform is needed.

Technology alone cannot solve the problem. Robust regulations and meaningful standards can ensure carbon credits reach their potential. Credits must go to projects that deliver genuine and verifiable emissions reductions. Assessing additionality (whether CO_2 would have declined without the project) more effectively, increasing the rigour of project audits, and actively involving communities in project development and implementation are all essential steps.

Plenty of money is available, and removing unnecessarily overpriced consultants will free up funds for the communities these projects are meant to benefit. Raise your hand if you agree! Shifting the view on carbon credits is vital. They are not just commodities to trade; they should be tools to drive real and lasting change. Moving away from the profit-driven model that runs the market and adopting an approach that values social and environmental outcomes is crucial for making a true impact.

The scandals and scams that have plagued the carbon credit market are a stark reminder that there are no easy fixes in the fight against climate change. The challenges we face are complex and multifaceted, and the solutions we develop need to be equally nuanced and robust. But the scandals and scams also offer a chance to learn, adapt and build a better system that delivers on the promise of carbon credits and helps to create a more sustainable and equitable world.

In the next chapter, we will explore transparency in the carbon market – what it means, why it matters and how we can achieve it. Transparency will be vital in fixing the flaws in the carbon credit system.

CHAPTER 5

THE QUEST FOR
TRANSPARENCY

CARBON CREDITS ARE CRAP

Transparency is often touted as the magic bullet in carbon credits. It will fix the market's flaws, restore trust and ensure that carbon offsets genuinely deliver on their promises. But what does transparency mean in this context? And how do we achieve it in a system that is complex and opaque by its very nature?

In its simplest form, transparency means that information is open, accessible and verifiable. In the carbon credit market, this translates to a system where every process step – from creating a carbon credit to its eventual purchase and use – is documented and available for scrutiny. It is about tracing each credit back to its source, understanding the impact of the project that generated it and knowing that the promised benefits have been delivered.

However, as we have seen, the reality of the carbon credit market is far from this ideal. Despite volumes of documents, glossy photos and fancy videos highlighting the alleged benefits of carbon credit projects, fraud and exploitation have found fertile ground. These well-crafted materials often serve more as marketing tools than evidence of genuine impact. They create a veneer of credibility that can be deeply misleading, allowing projects to appear legitimate while masking underlying issues.

The system is riddled with challenges that make true transparency difficult to achieve, from the complexities of monitoring large-scale projects in remote areas to the murky waters of additionality (proving that emissions reductions would not have happened without the project) and permanence (ensuring that these reductions last over time). When there is a lack of transparency, trust erodes and the entire market becomes vulnerable to fraud, exploitation and greenwashing.

THE PROBLEM WITH GLOSSY REPORTS AND FANCY VIDEOS

One of the most glaring issues with the current system is the reliance on glossy reports, slick marketing materials and well produced videos to convey the benefits of carbon credit projects. These materials often focus on the most visually appealing aspects of a project – pictures of lush forests, smiling children and state-of-the-art technology – while glossing over the less attractive details, such as the actual carbon reductions achieved, the socio-economic impacts on local communities, and the challenges faced in monitoring and verification.

These presentations can be persuasive, but they can also be misleading. They create an illusion of success that may not reflect the reality on the ground. Organizations that stand to benefit from the sale of carbon credits often produce these materials, compounding the problem. It creates a conflict of interest, where the incentive is to present the project in the best possible light rather than to provide an honest and transparent assessment of its impact.

Even with all these polished presentations, the underlying issues of fraud and misrepresentation persist. The Kariba REDD+ project, the Northern Kenya Rangelands Carbon Project and the cookstove carbon credit scandal – all of these cases had extensive documentation and promotional materials. Yet, as we now know, these materials did little to prevent the fraudulent activities that were taking place behind the scenes (see *Chapter 4*).

THREE WAYS TO INJECT TRANSPARENCY

The quest for transparency in the carbon credit market is not an easy one. It is like trying to find a vegan option at a Texas barbecue – possible, but it requires effort, creativity and a bit of scepticism. Transparency demands us to confront deep-seated issues that have allowed fraud and exploitation to flourish. It calls on us to adopt innovative technologies, simplify processes and fundamentally change how we think about transparency. But most of all, it requires a commitment to building a system that is as open, accountable and trustworthy as your grandma's recipe for chocolate chip cookies.

So, how do we inject transparency into a system that seems almost designed to resist it? The answer lies in a combination of technology, streamlined processes and a shift in mindset – each of which plays a crucial role in creating a more transparent and accountable carbon market.

THE ROLE OF TECHNOLOGY IN ENHANCING TRANSPARENCY

It is time to move beyond the surface-level transparency offered by glossy reports and videos. The carbon market needs to embrace technology that can provide accurate, verifiable data on the impact of carbon credit projects. Satellite imagery, remote sensing and AI-driven data analysis can offer objective insights into what is happening on the ground, far beyond what a photo or video can capture.

For example, satellite imagery can monitor deforestation in real time, providing indisputable evidence of whether a forest conservation project is preventing logging or other forms of degradation. Remote sensing can track the growth and health of vegetation, offering concrete data on carbon sequestration rates. AI-driven analysis can process vast amounts of data to detect patterns and anomalies that might indicate fraud or mismanagement.

These technologies can strip away the marketing gloss and reveal a project's actual state. They can help to ensure accurate, measurable reductions in greenhouse gas emissions to back up the carbon credits being sold. Moreover, they can provide the kind of continuous monitoring necessary to maintain transparency over the long term rather than relying on sporadic audits or self-reported data.

DIGITAL MONITORING, REPORTING AND VERIFICATION SYSTEMS

One of the most promising avenues for improving transparency in the carbon credit market is through advanced technologies, particularly digital monitoring, reporting and verification (DMRV) systems.[66] These systems leverage cutting-edge tech to track, monitor and verify carbon credits with the precision of a Swiss watch – if that watch also had a PhD in environmental science.

DMRV systems are the backbone of a transparent carbon credit market. Think of them as the hall monitors of the carbon credit world, ensuring everyone is where they should be and doing what they should be doing. These systems integrate digital tools to provide real-time performance data on carbon credit projects. By automating the collection and analysis of data, DMRV systems reduce the potential for human error and manipulation. Because, let's face it, humans are great at messing things up.

The need for such robust systems has become increasingly apparent as the carbon credit market has expanded. Traditional methods of monitoring and reporting, which often involved sporadic audits and manual data collection, are about as effective as using a butter knife to cut down a tree. As more companies and governments have committed to reducing their carbon footprints, the volume of data that needs to be monitored and verified has exploded. DMRV systems address this challenge by providing a scalable solution that can manage the complexity and volume of data required for effective carbon credit management. Precision and transparency are no longer optional, they are essential.

Moreover, the automation provided by DMRV systems ensures that data collection and analysis are consistent and objective, reducing the risk of human error or manipulation. Providing accurate and reliable data is crucial in a market where trust is paramount. DMRV systems deliver on this need by ensuring that every step of the carbon credit process is documented and verifiable.

SATELLITE MONITORING AND REMOTE SENSING

Satellite monitoring and remote sensing are potent tools that can significantly enhance transparency in the carbon credit market. These technologies allow for real-time tracking of environmental changes, such as forest cover or land use, providing up-to-date data that can verify the impact of carbon credit projects. It is like having Google Earth but with a conscience.

By integrating satellite data with on-the-ground reports, these tools offer a comprehensive view of a project's environmental impact, ensuring that the claimed emissions reductions are legitimate.

The use of satellite technology has revolutionized the way we monitor environmental changes. In the past, assessing the impact of carbon credit projects often required costly and time-consuming site visits – imagine trying to find a needle in a haystack, but the haystack is the size of Texas. Satellite monitoring has eliminated much of this burden by providing a constant stream of data that can be accessed from anywhere in the world. The technology reduces the cost of monitoring and increases the frequency and accuracy of data collection.

Remote sensing technology also plays a critical role in ensuring the permanence of carbon credits. One of the biggest challenges in the carbon market is ensuring that the emissions reductions achieved by a project are permanent and not reversed by future events such as deforestation, wildfires or land use changes. Satellite monitoring provides early warnings of these threats, enabling swift action to protect the carbon savings achieved. Such monitoring is particularly important for projects involving reforestation or afforestation (planting trees in an area that didn't previously have them), as the long-term success of these initiatives hinges on the continued health and growth of the trees.

ARTIFICIAL INTELLIGENCE AND DATA INTEGRATION

Unbelievably, measuring carbon absorption today in forests relies on a surprisingly low-tech approach. Field workers head into the forest with only a tape measure and a clipboard. Yes, really – check out the YouTube videos if you don't believe me.[67] These resolute workers measure tree trunks, jot down the numbers and move on to the next tree. But let's be honest – there is no way they can measure every tree across a million-hectare area.

Artificial intelligence (AI) is revolutionizing the carbon credit market by providing precision, transparency and fraud prevention in one powerful package. Relying on AI instead of manual measurements enables the analysis of massive datasets from satellite imagery, LiDAR (a method for visualizing the Earth's surface), ground sensors and other remote sensing technologies. Such technology allows for accurate calculations of the carbon absorption of individual trees and entire forests. Swapping a handheld flashlight for a satellite view offers a comprehensive perspective on every tree in the forest.

AI does not stop at measuring tree trunks; it estimates the carbon absorption capacity of each tree based on its species, size and health. By cross-referencing data from various sources, AI provides a precise estimate of the emissions reductions a project achieves – like a detective piecing together clues to solve the case of the missing carbon. This approach ensures that carbon credits genuinely reflect the environmental benefits achieved.

But AI's capabilities go even further. It acts as a vigilant watchdog, detecting anomalies and inconsistencies in the data and flagging potential issues before they escalate. For example, AI can alert the relevant stakeholders if a project reports emissions reductions significantly higher than expected based on similar projects. This early intervention helps to prevent fraud and misreporting.

Data integration enhances transparency in the carbon credit market. By linking data from satellite imagery, ground sensors and project reports, AI creates a comprehensive and up-to-date picture of each carbon credit project. This approach improves the accuracy and reliability of the information, making it easier to detect and address discrepancies or issues.

Proper data structuring is also critical. By organizing and standardizing the data collected by DMRV systems, satellite monitoring and AI analysis, we can ensure that the information is accessible, comparable and verifiable. We can adopt uniform data standards across the industry, ensuring that all relevant information, such as project location, methodology, emissions reductions and community impact, is consistently recorded. A well-organized data set is satisfying and efficient.

With this level of data structuring and integration, we can create a centralized database with traceable accounts for all carbon credit transactions and project details. This database can be accessible to all stakeholders, including project developers, regulators, investors and the public, ensuring the entire carbon credit lifecycle remains transparent and traceable.

AI and data integration can transform the carbon credit market from a rough, manual process into a sophisticated, transparent and efficient system. AI is paving the way for a more reliable and trustworthy carbon credit market by bringing precision to carbon accounting, enhancing transparency and preventing fraud.

STREAMLINING
THE CARBON CREDIT
ECOSYSTEM

Another key to enhancing transparency is streamlining carbon credits' creation, verification and sale. The current system is bogged down by layers of bureaucracy, complex verification procedures and a web of intermediaries, which creates opportunities for fraud and makes it difficult to track the flow of credits from creation to final use.

Simplifying these processes can reduce the chances of manipulation and make it easier to hold all parties accountable. For instance, standardizing the criteria for additionality and permanence across projects could help to ensure that all credits meet a consistent quality baseline. Reducing the number of intermediaries involved in the sale of credits could make the system more transparent and reduce the opportunities for profit-driven manipulation.

USING TECHNOLOGY
OR STREAMLINING

The advanced technologies mentioned above – DMRV systems, satellite monitoring and AI – reduce the need for traditional standards boards and other intermediaries. Initially established for oversight, these organizations often face criticism for inefficiency, lack of transparency, and vulnerability to corruption. As Chairman and CEO of a technology company in Africa, I aim to eliminate traditional standards boards. Having had 30 years to prove their worth, they should face termination. Shifting toward digital tools and direct reporting reduces bureaucracy and creates a more agile, transparent market.

The carbon credit ecosystem involves eight key players, comprising over 100,000 companies. Each player has their hand in the carbon credit cookie jar long before any so-called community benefits come into play. Wouldn't it be great to rid ourselves of some of these hungry hippos?

- **Investors**: Investors sponsor and finance carbon credit projects, providing the capital to launch them.
- **Project developers**: Developers design and run the carbon offset projects and sell the resulting carbon credits to buyers.
- **Carbon registries**: These organizations validate and verify carbon credit projects. They independently assess emissions reduction potential before registering projects and regularly monitor actual reductions once operational, supposedly ensuring ongoing compliance and accuracy.

- **Carbon offset programme schemes**: These set standards for carbon credit quality, certify and issue carbon credits, and maintain a registry to track certified projects and credit issuance.
- **Carbon exchanges**: Exchanges are marketplaces where project developers list verified carbon credits for buying and selling, making trading these credits easier.
- **Carbon credit rating agencies**: Rating agencies use their data and criteria to assess projects. They evaluate the quality and credibility of carbon credit projects, adding another layer of analysis – some might say data overkill.
- **Carbon brokers/retailers**: These companies simplify the process for buyers, managing the complexities of selecting and purchasing credits and making transactions quicker and more efficient – although it is questionable whether this simplification service is worth the fees the companies charge.
- **Credit buyers**: Companies seeking to offset their carbon footprint are often the end purchasers of carbon credits.

Is your head spinning yet? With all these players involved, it is easy to see how the process can become a tangled mess. In the traditional model of the carbon market, these players operate in silos, leading to inefficiencies, increased costs and a lack of transparency. Carbon credits frequently pass through multiple intermediaries before reaching the end buyer, with each intermediary adding complexity and costs.

Despite the existence of all these players, carbon credits are still crap. My goal is to eliminate all unnecessary intermediaries in this ecosystem. Simplifying the process and cutting out intermediaries creates a more efficient and transparent market – because who doesn't love cutting out the middleman?

Data silos are detrimental to the success of any business or organization, and the carbon credit ecosystem is no exception. With over 100,000 stakeholders working independently, chaos and confusion have taken root, hindering the effectiveness and transparency of the market. Breaking down these silos and fostering data integration creates a more cohesive and efficient system that benefits everyone involved (*Chapter 6* will explore this topic in more detail).

We also need to empower countries to manage their natural resources and carbon credits and digitize them, putting proper guardrails in place. Local governments and communities must have control over their assets while maintaining transparency and accountability. Strengthening local ownership aligns incentives with environmental protection and sustainable development.

Streamlining the ecosystem would offer numerous benefits. It would reduce the cost of carbon credits, making them accessible to a broader range of buyers. This is crucial for small and medium-sized enterprises, which may lack the resources to navigate the complex and costly traditional market. Saving the planet should not break the bank.

Streamlining would also boost the speed and efficiency of transactions. In the traditional model, carbon credits can take years to move from project inception to final sale – like waiting for a sloth to cross the US. By reducing intermediaries and automating much of the process, we can significantly shorten this timeline, allowing for quicker deployment of carbon credits. According to the climate 'experts,' every second counts, so let's act with the urgency that this challenge demands.

REVOLUTIONIZING THE CARBON CREDIT ECOSYSTEM WITH FIÙTUR

The most critical change needed to achieve true transparency in the carbon market is a shift in mindset. The focus needs to move away from marketing and toward accountability. This means prioritizing the accurate and honest reporting of project impacts over the creation of visually appealing materials. It means embracing a culture of openness, where data is shared freely, and stakeholders are encouraged to ask challenging questions.

It also means recognizing that transparency is not just a tool for preventing fraud but a fundamental component of a functioning carbon market. Without transparency, there can be no trust. And without faith, the entire system collapses. Let's examine the mindset shift through an extended example.

JOE MADDEN:
LEADING THE CHARGE
INTO TOMORROW

Meeting Joe Madden, CEO of Fiùtur,[68] it is clear he is significantly impacting environmental technology. As a visionary entrepreneur, Joe has profoundly contributed to the nexus of markets, data and environmental impact.

With Joe Madden at the Dream On Gala,
supporting Promises2Kids, in 2024

We discussed the fragmented carbon credit market and brainstormed innovative solutions. Our day concluded at a charity event for foster care kids – a cause close to our hearts – where we enjoyed dancing to Kool & the Gang. This underscored the importance of community and collective impact.

In 2015, Joe co-authored a transformative paper on leveraging commodity production data to enhance market value assessments.[69] He then co-founded Xpansiv with the mission of digitizing global commodity markets.

Xpansiv initially focused on revolutionizing the digitization of natural gas for financial institutions, introducing a more data-driven and efficient trading environment. The initiative set the stage for what would become a broader environmental impact. Xpansiv emerged as a leader in creating transparent, market-driven responses to urgent climate issues by quantifying environmental attributes in commodity production and trading them on a digital platform.

The pioneering approach enabled a significant leap forward for the environmental commodities market. Xpansiv merged with CBL Markets to form a comprehensive platform called Intelligent Commodities™. The platform became the first to integrate 24 carbon registries and offer access to over 200 environmental spot commodities in a unified market.

Before his time at Xpansiv, Joe co-founded EOS Climate, which played a pivotal role in California's greenhouse gas markets and earned a prestigious Governor's Environmental and Economic Leadership Award (GEELA), the state's highest environmental honour.

Building on the thriving digital strategies employed by Xpansiv, Fiùtur, a spin-off, stands as a beacon of innovation in the carbon credit sector. Fiùtur uses the digital infrastructure developed by Xpansiv to revolutionize carbon credit management with its SMART system, ensuring transparency, efficiency and accountability, and making each credit as impactful as it promises.

Joe passionately explained how Fiùtur's SMART system revolutionizes the management and verification of carbon credits by integrating cutting-edge digital technologies.

Like a precision-engineered machine, it streamlines the entire lifecycle of carbon credits with substantial environmental benefits.

OVERVIEW OF FIÙTUR'S SMART SYSTEM

Fiùtur's SMART system epitomizes a comprehensive approach to carbon credit management.[70] An apt acronym for 'sustainable monitoring, assessment, reporting and tracking,' it ensures verifiable data back up each carbon credit. The system provides a holistic view of carbon projects, enabling stakeholders to monitor progress and adhere to environmental and social standards through advanced technologies such as satellite monitoring, AI-driven data analysis and automated reporting.

The SMART system's architecture is modular and scalable, like a Swiss Army knife for environmental management. It allows for integrating emerging technologies and has positioned Fiùtur as a leader in carbon management.

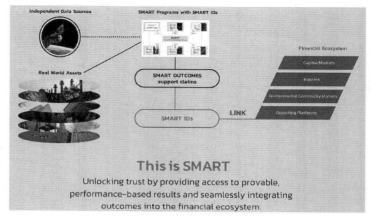

Fiùtur's SMART system workflow

The SMART system has four key characteristics:

- **Continuous real-time monitoring**: A standout feature of SMART is its real-time monitoring capability, which instantly identifies and rectifies any discrepancies. This ensures the integrity of carbon credits through a continuous influx of data from satellites, ground sensors and Internet of Things devices.
- **Integration of advanced technologies**: By merging satellite imagery with AI and data analytics, SMART offers unmatched precision and detail in project monitoring. Such integration reduces the need for costly on-site verification, streamlining the process and making carbon credits more accessible and affordable.
- **Transparency and Accessibility**: Designed for ease of use, SMART allows all stakeholders to access and understand project data regardless of their technical prowess. Its user-friendly interface and robust data

visualization tools simplify navigating and interpreting information, fostering a transparent and accountable carbon market.

- **Fostering a shift in mindset**: SMART aims to transform the carbon market by promoting a shift from profit-centric motives to prioritizing social and environmental impacts. It encourages trust and collaboration, which are essential for a transparent and effective carbon market.

THE PATH FORWARD

The pursuit of transparency in the carbon market is a shared effort among all involved parties. As several companies, including Fiùtur, enter the space with similar ambitions, the landscape is becoming highly competitive. Fiùtur, along with others like Senken[71] and KlimaDAO,[72] deploy technology to navigate and simplify market complexities, transforming challenges into avenues for innovation. With the market evolving, it's too soon to determine which company will emerge as the leader in setting technology standards for a sustainable and transparent carbon market.

In the next chapter, we will explore emerging innovative approaches to carbon credits that emphasize community involvement and creative solutions, setting the stage for the future of the carbon market. Doing full carbon credit transparency right means making it sustainable, equitable and fun. Who doesn't appreciate a little disruption now and then?

CHAPTER 6

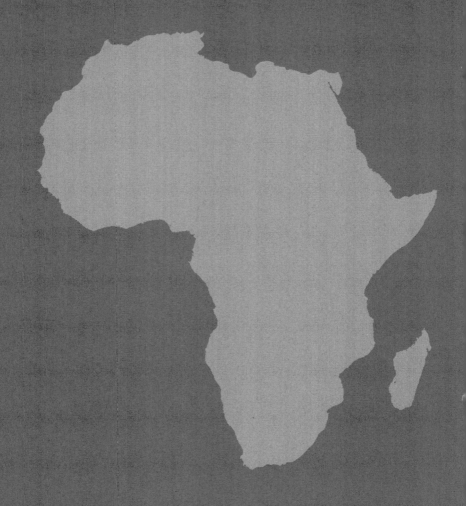

INNOVATING THE CARBON CREDIT MARKET

As we have explored the complexities, pitfalls and potential of the carbon credit market, one thing has become abundantly clear: the system needs a significant overhaul. I know, I know, I sound like a broken record. Yet, amid the challenges, there is a growing wave of innovation – original approaches that aim to address the shortcomings of the traditional carbon credit market while driving meaningful, sustainable change.

In this chapter, we will dive into some of the most promising innovations in the carbon credit market. From community-driven models to innovative technological solutions and groundbreaking carbon sequestration projects, these approaches are paving the way for a future where carbon credits are more than just financial instruments – they are fundamental catalysts for positive change. And let's be honest – if we are serious about fixing the broken system, it is time we did it right. After 30 years of watching what the so-called experts have accomplished, we need an innovative approach.

COMMUNITY-CENTRIC CARBON CREDITS

One of the traditional carbon credit market's most significant criticisms is its tendency to overlook the needs and voices of the local communities most affected by carbon offset projects. Too often, these communities are seen as passive beneficiaries rather than active participants in the development and implementation of projects. The disconnect has led to projects failing to deliver the promised benefits to those needing them most.

But what if we flipped the script? Instead of imposing top-down solutions, we could build carbon credit projects from the ground up, with local communities at the centre. Say what? This is the idea behind community-centric carbon credits – a model that emphasizes local ownership, control and benefit-sharing. In this approach, carbon credit projects are designed and implemented in close collaboration with local communities, ensuring that their needs, knowledge and perspectives are fully integrated into the project.

LOCAL OWNERSHIP AND CONTROL

At the heart of community-centric carbon credit projects is the principle of local ownership and control. Rather than approaches being dictated by external developers or investors, the communities manage these projects themselves, with support from technical experts and financial backers. Communities are empowered to take charge of the projects' development, ensuring that the benefits – whether in revenue, jobs or environmental improvements – remain within the community.

With Nansana Municipal Council, Uganda

In my proposal, communities engage in every project stage, from planning and design to implementation and monitoring. They influence the project structure, prioritize activities and help to decide how to distribute the profits. The proposed approach increases the likelihood of the project's success and builds local capacity and resilience, helping communities to better manage their natural resources and adapt to the impacts of climate change.

BENEFIT-SHARING MECHANISMS

A key component of community-centric carbon credits is developing robust benefit-sharing mechanisms. These mechanisms ensure that community members equitably receive the financial rewards generated by carbon credits, preventing external actors from siphoning them off. Benefit-sharing can take many forms, depending on the community's needs and priorities. In some cases, it might involve direct cash payments to community members. In contrast, in others, the funds might be used to support local development projects, such as building schools, improving healthcare or providing clean water. The important thing is that the community has a say in using the benefits, ensuring they align with local priorities and contribute to long-term sustainable development.

INTEGRATING TRADITIONAL KNOWLEDGE

Another critical aspect of community-centric carbon credit projects is integrating traditional knowledge and practices. Many Indigenous and local communities deeply understand their natural environment, which has developed over generations and centuries. Their knowledge is invaluable in designing and implementing carbon credit projects that are effective, culturally appropriate and sustainable.

For example, traditional agricultural practices, such as agroforestry and rotational grazing, can be crucial in carbon sequestration, enhancing biodiversity and improving soil health. By incorporating these practices into carbon credit projects, communities can build on their strengths and create innovative solutions rooted in local traditions.

ADDRESSING ADDITIONALITY WITH INNOVATION

One of the most challenging issues in the carbon credit market is additionality – ensuring that carbon credits represent genuine additional emissions reductions. While this issue has been a persistent source of controversy in the market, a recent innovation offers new ways to tackle it head-on.

DYNAMIC BASELINES

Traditional carbon credit methodologies use static baselines founded on historical data and assume that emissions will stay constant without intervention. However, this approach is problematic, as it does not account for changes in economic conditions, technology or policy that could affect emissions levels in the future.

In contrast, dynamic baselines receive regular updates to reflect current conditions, making it possible to more accurately assess whether a carbon credit project is truly additional – that is, whether it delivers emissions reductions that would not have occurred without the project. Using dynamic baselines, we ensure that carbon credits represent tangible, measurable and additional contributions to climate change mitigation.

ADDRESSING PERMANENCE WITH INNOVATION

In addition, permanence has attracted significant controversy in the carbon credit market. Ensuring the permanence of emissions reductions poses a critical challenge. A forest that sequesters carbon today might be cut down or burned tomorrow, releasing that carbon back into the atmosphere. Innovators are exploring various mechanisms to guarantee the permanence of carbon credits.

INSURANCE MECHANISMS

Under this model, carbon credit projects must buy insurance policies that cover non-permanence risks. Suppose a project fails to maintain its emissions due to deforestation, natural disasters or mismanagement. In that case, the insurance policy compensates by providing funds to restore the carbon savings or buying additional credits to offset the loss.

This approach not only offers a financial safety net for carbon credit buyers but also motivates project developers to proactively take measures to protect the long-term integrity of their projects. By aligning financial incentives with environmental outcomes, insurance mechanisms help ensure carbon credits' permanence.

STRATEGIES FOR PROJECT DEVELOPERS

Developers can incorporate a range of strategies to further bolster the permanence of their projects:

- **Buffer pools**: Establish buffer reserves of carbon credits, which will act as insurance to cover losses from unforeseen events that may compromise a project's carbon sequestration capabilities. Buffer pools do exist, but unfortunately they were not large enough to cover the fraudulent carbon credits issued in the projects explored in *Chapter 4*.
- **Long-term land leases or ownership**: Secure long-term control over project land through leases or purchases to safeguard the carbon storage site from alternative uses that could compromise its integrity.
- **Legal and regulatory frameworks**: Collaborate with government bodies to develop legal protections that enforce the permanence of carbon sequestration, such as designating lands as permanent conservation areas.
- **Community management agreements**: Engage local communities by forming agreements that involve them

directly in the management and benefit-sharing of carbon projects, thus ensuring local investment in the project's success and longevity.

- **Permanence-monitoring technology**: Employ advanced technologies, such as satellite imagery and drone surveillance, to continuously monitor the health and stability of carbon-sequestering areas.
- **Reinforcement planting**: Implement strategies such as reinforcement planting to compensate for potential losses over time, ensuring forest density and sequestration capacity are maintained or increased.
- **Staggered project timelines**: Develop projects with staggered timelines to allow management of new areas as older ones mature, creating a sustainable carbon sequestration cycle.
- **Permanence audits**: Conduct regular independent audits to verify the continued effectiveness of sequestration measures and the integrity of the carbon credits.

RIPPLENAMI'S CARBON CLARITY: A NATIONAL-LEVEL SOLUTION

While community-centric models and innovations addressing additionality and permanence are crucial, the chaotic landscape of carbon credit management also demands national-level solutions. As the chairman and CEO of RippleNami,[71] I am excited to discuss how the company's Carbon Clarity™ technology is making a difference.[72]

For over a decade, RippleNami has led the use of technology to boost sustainable growth and lessen dependency on aid in developing regions, focusing exclusively on Africa. The continent is awash with rich sources of data, offering unique opportunities to solve critical economic challenges. Initially created to manage emergency chaos, RippleNami developed scalable, cloud-hosted web and mobile applications that provide real-time, actionable information. These tools facilitate rapid communication and coordination in critical situations.

Building on this robust foundation, RippleNami has pivoted to focus on its traceability and transparency platform, using AI and data visualization to consolidate and streamline data from various government ministries and agencies. By dismantling data silos and enhancing collaboration across different sectors, RippleNami empowers governments to address persistent challenges like uncollected taxes, inefficient resource allocation and limited access to essential services. The company's solutions ensure information is accessible, verifiable and actionable, equipping stakeholders with the detailed, comprehensive data they need to make well-informed decisions.

A significant impact area for RippleNami is tax compliance. The company has developed and deployed advanced tax compliance systems that help governments tackle the challenge of uncollected taxes in areas from rental income to property and other sources. These systems refine tax collection processes and boost economic equity and governance by collecting revenue efficiently and transparently.

Beyond tax compliance, RippleNami has played a vital role in connecting communities with essential services. Through partnerships with non-governmental and other organizations, RippleNami has mapped and integrated information on healthcare, education, agriculture, mining, transportation, infrastructure and financing, making this data readily available to those who need it most. Our comprehensive approach empowers local communities by providing them with the information necessary to make informed decisions and enhance their livelihoods.

RippleNami has been instrumental in developing real-time identification and traceability programmes in the agricultural sector. These programmes improve the management of resources such as livestock and produce, tracking them from origin to consumer. This transparency enhances food safety and quality and strengthens the supply chain, benefiting producers and consumers.

RippleNami's technology addresses global challenges like food security and the impacts of pandemics. By developing digital survey applications and other tools, RippleNami equips governments and organizations with the data to assess and respond to crises, ensuring effective resource allocation.

In the rapidly evolving field of carbon credits, we focused on studying the country-level technology and tools essential for monitoring the climate economy. We discovered that all countries lacked the necessary tools and structures. RippleNami's Carbon Clarity platform stands out as a beacon of transparency. The platform aids in the registration, compliance and management of carbon credit projects, ensuring they deliver tangible environmental and community benefits. By providing a clear, accessible view of how carbon credits are generated and used, RippleNami fosters trust and accountability in the market.

Through its diverse projects, RippleNami has proven that data transparency is a potent tool for driving real change, not just a buzzword. By making data accessible and actionable, RippleNami empowers governments, organizations and communities to address complex challenges, enhance governance and forge sustainable, long-term solutions.

Our innovative approach is transforming the landscape of carbon credit management, especially in Africa. If there's one lesson I've learned, innovation is not just about creating fancy gadgets – it's about solving real problems that impact real people.

THE PROBLEM

The current state of carbon credit management, particularly in Africa, is plagued by fragmentation. With over 100,000 carbon credit stakeholders operating independently (see *Chapter 5*), data silos have emerged, creating a disjointed system that hampers transparency, traceability and accountability. Data silos are detrimental to the success of any business or organization.

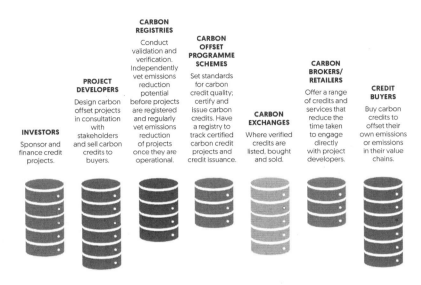

CARBON REGISTRIES
Conduct validation and verification. Independently vet emissions reduction potential before projects are registered and regularly vet emissions reduction of projects once they are operational.

PROJECT DEVELOPERS
Design carbon offset projects in consultation with stakeholders and sell carbon credits to buyers.

INVESTORS
Sponsor and finance credit projects.

CARBON OFFSET PROGRAMME SCHEMES
Set standards for carbon credit quality; certify and issue carbon credits. Have a registry to track certified carbon credit projects and credit issuance.

CARBON EXCHANGES
Where verified credits are listed, bought and sold.

CARBON BROKERS/ RETAILERS
Offer a range of credits and services that reduce the time taken to engage directly with project developers.

CREDIT BUYERS
Buy carbon credits to offset their own emissions or emissions in their value chains.

The chaotic climate ecosystem

The lack of coordination is detrimental to nations, enterprises and organizations. Worse still, all countries lack a national-level carbon credit traceability, monitoring, reporting, auditing and taxation programme, which is essential to effectively managing the complexities of carbon credits.

Take the Kariba REDD+ project as an example. As described in *Chapter 4*, this project has faced significant issues with transparency. The promised €30 million in co-share benefits never reached the intended communities, and the tax revenue generated from the project's carbon credit sales did not benefit the country's leaders or its people. This scenario highlights the urgent need for a more cohesive and transparent approach to carbon credit management.

THE CASE FOR TRANSPARENCY: A SOLUTION

Carbon Clarity is RippleNami's answer to this chaos. The technology eliminates data silos by integrating information from various stakeholders – including investors, developers, certifiers, exchanges, brokers, resellers, buyers, co-benefit recipients and tax authorities – into a single comprehensive profile. Doing so empowers countries like Zimbabwe (the location of the Kariba REDD+ project) to manage their carbon credit data more effectively, ensuring compliance, transparency and revenue collection. And let's be honest, who doesn't like the sound of less paperwork and more actual results?

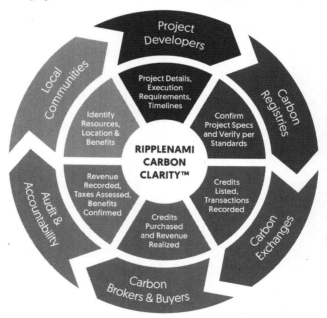

RippleNami's Carbon Clarity™ system

Carbon Clarity streamlines carbon credit management at the national level by organizing data into six distinct modules:

- The **Project Developers** module tracks and manages data from entities responsible for creating and maintaining carbon credit projects. It includes all their in-country registrations and relevant documents from outside the country, ensuring their activities remain transparent and accountable.
- The **Carbon Registries** module integrates data from carbon registries, providing a centralized record of all issued credits and simplifying the process of tracking their lifecycle.
- The **Carbon Exchanges** module monitors the trading of carbon credits on various exchanges, ensuring that transactions are transparent and that all parties are held accountable.
- The **Carbon Brokers & Buyers** module tracks buyers and informs them about the projects they support and purchase credits from.
- The **Audit & Accountability** module offers real-time monitoring and verification tools, allowing for continuous oversight and reducing the risk of fraud.
- The **Local Communities** module ensures that carbon credit projects' benefits reach local communities as promised, tracking the distribution of funds and other benefits to align with community needs and priorities.

By integrating these modules, Carbon Clarity provides a comprehensive, transparent and efficient system for managing carbon credits at the national level. This modular approach prevents issues like those seen in projects such as Kariba REDD+ and ensures that benefits are distributed fairly, supporting sustainable development in local communities.

LEVERAGING TECHNOLOGY FOR PRECISION AND ACCOUNTABILITY

As we have discussed, technology is critical in improving transparency and accountability in the carbon credit market. But beyond monitoring and reporting, technology can also enhance the precision and effectiveness of carbon credit projects, ensuring that they deliver the maximum possible impact.

PRECISION AGRICULTURE AND CARBON FARMING

One area where technology is making a significant difference is in precision agriculture and carbon farming. By using advanced tools such as drones, sensors and satellite imagery, farmers can optimize their land management practices to maximize carbon sequestration while improving crop yields and reducing inputs like water and fertilizer.

Precision agriculture allows for more accurate measurement of carbon sequestration in soils and vegetation, providing a solid foundation for generating carbon credits. At the same time, it can help farmers adopt sustainable practices, such as no-till farming, cover cropping or agroforestry, which sequester carbon and enhance soil health, biodiversity and resilience to climate change.

AI-DRIVEN PROJECT DESIGN AND OPTIMIZATION

AI can be used to design and optimize carbon credit projects, ensuring they deliver the most significant impact. By analysing large datasets from past projects, AI can identify the most effective strategies for reducing emissions and sequestering carbon in different contexts.

For example, AI can determine the optimal mix of tree species for reforestation projects, considering growth rates, carbon storage potential, and resilience to pests and diseases. It can also model the potential impacts of different land management practices, helping project developers choose the approaches that will generate the most carbon credits and thereby maximize the project's return.

AI-driven optimization can also enhance the financial performance of carbon credit projects, identifying opportunities to reduce costs, increase efficiency and maximize revenue. This makes it easier for project developers to create economically viable projects with strong environmental and social outcomes.

While technology plays a crucial role in optimizing carbon credit projects, the purest and most direct form of carbon capture remains rooted in nature. In the next chapter, we explore how natural carbon sinks – such as forests, bamboo, biochar and ocean ecosystems – offer unmatched potential for removing carbon from the atmosphere. We will examine the science behind these natural solutions, their challenges and how they can be integrated into a comprehensive climate strategy.

CHAPTER 7

THE PUREST FORM OF CARBON CAPTURE

A DEEP DIVE INTO CLIMATE SCIENCE WITH DR WILLIAM DEWAR

One of the most profound moments in my journey to understand carbon capture came when I was honoured to speak to Dr William Dewar. Before discussing our conversation, let me provide you with Dr Dewar's impressive credentials.

Dr Dewar's academic journey began with a BSc in physics (with a minor in mathematics) from Ohio State University in 1977. As a die-hard Michigan football fan, I tried not to hold the Ohio State degree against Dr Dewar – though I must admit, it crossed my mind. Go Blue!

Dr Dewar furthered his studies at MIT, earning an MSc in physical oceanography in 1980 and a PhD in 1983. He has dedicated his career to studying the complex interactions between the ocean and the atmosphere, which are fundamental to regulating the Earth's climate.

Now, raise your hand if you have ever heard of physical oceanography. Anyone? Anyone? Don't worry – I hadn't either until I met Dr Dewar. But it turns out physical oceanography is vital for climate studies. The ocean is not just a big blue thing we sail on – it is a significant player in the Earth's climate system. It absorbs a massive amount of the sun's heat and CO_2 from the atmosphere, influencing weather patterns, sea levels and even where marine life decides to hang out.

Dr Dewar's work has been vital to understanding how all these processes operate, which is essential for predicting future climate change and developing strategies to mitigate its effects. Meanwhile, I am just a simple Texas girl with an accounting degree from Texas Tech University, where we say simple things like "Get Your Guns Up." I am trying to wrap my head around all this science talk. Talking with Dr Dewar was like trying to keep up with a rocket scientist – only this rocket was diving deep into the climate, and I was trying to stay grounded.

Dr Dewar's expertise in oceanic and atmospheric interactions provides critical insights into the broader field of climate science. He emphasizes the importance of understanding our oceans to fully grasp global climate dynamics' complexities. His research advances scientific knowledge and informs practical approaches to addressing the challenges posed by climate change, making his contributions invaluable to the scientific community and the world.

During our insightful conversation, Dr Dewar shared a simple yet compelling thought: "Jaye, the purest form of carbon capture is to plant something, let it grow and then bury it." This idea resonated with me, and I mentioned the potential of hemp and biochar as simple yet powerful methods for carbon capture. As we discussed further, it became clear that hemp could play a significant role in capturing carbon and offer numerous environmental and economic benefits, especially when paired with biochar.

INDUSTRIAL HEMP'S DIRTY LITTLE SECRETS

I need to digress for a quick moment. Did you know industrial hemp existed for thousands of years before it was catapulted into our political and economic consciousness in the 20th century? As one of the earliest plants to be cultivated by humanity, hemp has a rich and varied history, woven into the fabric of countless civilizations. Ancient China, the Roman Empire, and many other cultures relied on hemp for textiles, paper, food and medicine.

Hemp's durability and versatility made it indispensable. The ancient Chinese were among the first to recognize its potential, using hemp fibres to create some of the earliest paper around 150 BCE. Strong fibres also produced ropes, sails and clothing, making them an essential resource for centuries.

Fast-forward to colonial America, where hemp held such value that farmers faced legal requirements to grow it. Early drafts of the Declaration of Independence were written on hemp paper. For generations, hemp was the backbone of many industries, particularly textiles, ropes and sails. Without hemp, ships exploring and connecting the world might never have sailed.

So, what happened? How did a plant integral to human progress become vilified and banned in the US? Enter the 20th century, where industrial interests began shaping public policy more than ever.

By the 1930s, industrialists like John D. Rockefeller, with his vast petroleum empire, and William Randolph Hearst, with his massive timber and paper holdings, saw hemp as a significant threat. Hemp could produce bio-fuels and cheaper, more sustainable paper, directly challenging these industrial titans' businesses. They had the money and influence to protect their interests, and they did so ruthlessly.

Rockefeller and Hearst, along with chemical giant DuPont – which had just developed nylon, a synthetic fibre that could replace hemp – lobbied heavily against hemp. They painted it with the same brush as marijuana, stoking racial fears and exploiting the public's ignorance about the differences between industrial hemp and its psychoactive cousin.

Hearst, in particular, used his media empire to wage a propaganda campaign against hemp, linking it to the 'dangers' of marijuana and crime, often with racial undertones aimed at Mexican immigrants and African Americans. The fearmongering worked. In 1937, the US Congress passed the Marihuana Tax Act, which banned hemp by making it prohibitively expensive to produce.

For decades, the ban on hemp stifled innovation and suppressed a crop that had been a staple of human industry for millennia. Not until the 2018 Farm Bill,[73] which finally differentiated industrial hemp from marijuana, could hemp make its legal return to American agriculture.

Hemp's story is not just about plants. It is about how economic interests and political power reshape history, suppressing an industry that could have continued to thrive for the benefit of many. The ban on hemp serves as a cautionary tale of what happens when corporate interests dictate public policy and reminds us that progress is not always a straight line.[74]

NATURE'S CARBON SUPERSTARS: WEED AND DIRT?

Let's clear the air (and the smoke) before we dive into this topic. As explored above, we are talking about industrial hemp here – not the kind that makes you ponder the mysteries of the universe while munching on an endless supply of snacks. Industrial hemp is all business.

TEXTILES

– Fabrics
– Handbags
– Shoes
– Diapers
– Denim
– Clothing

INDUSTRIAL TEXTILES

– Rope
– Canvas
– Tarps
– Carpeting
– Netting
– Caulking

PAPER

– Printing
– Newsprint
– Cardboard
– Packaging

BUILDING MATERIALS

– Bricks
– Fibreboard
– Insulation
– Acrylics
– Hempcrete

FOODS

– Hemp seeds
– Seed oil
– Protein powder
– Supplements

INDUSTRIAL PRODUCTS

– Oil paints
– Varnishes
– Printing inks
– Fuel
– Solvents
– Coatings

BODY CARE

– Soaps
– Lotions
– Balms
– Cosmetics
– Shampoos

LEAVES

– Mulch
– Compost
– Animal bedding

HEALTH

– Pain relief
– Skin health
– Anti-inflammatory
– Detoxification
– Nutritional content

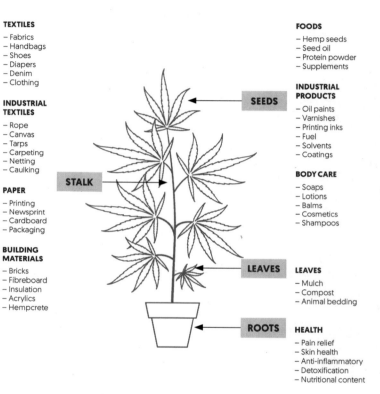

SEEDS

STALK

LEAVES

ROOTS

The many uses of hemp

With its deep root system, industrial hemp is like the overachieving cousin of the hemp family – digging deep to improve the soil structure, enhance water infiltration and reduce erosion. It is not the chill-out kind of hemp but the get-your-hands-dirty-and-save-the-planet kind. Industrial hemp is an unstoppable soil superhero. It is an expert in phytoremediation (big word even for spellcheck), which can absorb heavy metals and pollutants from the soil, like cleaning up after a wild party. And if that isn't impressive enough, it has even been planted around the Chernobyl nuclear disaster site to help remove radioactive elements from the soil.[75] So, while marijuana-type hemp might help you unwind, industrial hemp is out there saving the world – one contaminated field at a time.

Hemp is not just a one-trick pony – it has significant benefits. With about 50,000 different uses, it is more like duct tape (though, honestly, I have not verified the 50,000 uses). Beyond its superhero status in soil health, hemp's rapid growth cycle and minimal resource needs make it a dream crop for farmers, especially in areas where water and fertilizers are scarce. But its impact doesn't stop there.

Hemp by-products, from nutritious food products and sustainable textiles to biodegradable plastics and eco-friendly construction materials, have the potential to create jobs, boost local economies and even save the planet. No cape is required – just inspiration and motivation.

BIOCHAR

When you add biochar to the mix, hemp's benefits go from impressive to magical. Biochar is charcoal's eco-friendly cousin. It is made from plant waste sources, such as trees, feed crops and hemp.

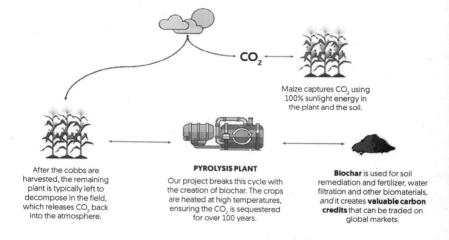

CO$_2$

Maize captures CO$_2$ using 100% sunlight energy in the plant and the soil.

After the cobbs are harvested, the remaining plant is typically left to decompose in the field, which releases CO$_2$ back into the atmosphere.

PYROLYSIS PLANT
Our project breaks this cycle with the creation of biochar. The crops are heated at high temperatures, ensuring the CO$_2$ is sequestered for over 100 years.

Biochar is used for soil remediation and fertilizer, water filtration and other biomaterials, *and* it creates **valuable carbon credits** that can be traded on global markets.

Caelum's crop waste biochar process

Biochar is a real overachiever – it can be buried in the soil and sequester carbon for centuries. So, while hemp captures annoying CO$_2$ from the atmosphere, biochar locks CO$_2$ down and puts it back into the soil, improving its structure, boosting water retention and giving a cosy home to all the good little microorganisms. Plus, biochar can reduce the need for chemical fertilizers, making your garden greener in many ways. It is like turning dirt into gold – if gold saved the planet.

My conversation with Dr Dewar highlighted a powerful approach to mitigating climate change. By integrating hemp cultivation with biochar production, we can create a comprehensive, full-circle solution for carbon capture. This approach supports economic development, enhances soil health and fosters environmental restoration, but it also empowers us to be part of the solution to one of the most pressing issues of our time.

To gain a deeper understanding of biochar, I spoke with Tom Miles, a leading expert in biochar, who has dedicated much of his career to advancing biochar technologies and applications. Our conversation provided valuable insights into the potential of biochar as a critical tool in carbon sequestration.

At Oregon Biochar Solutions with Karl Strahl, chief operating officer

Through Tom's guidance, I landed at Oregon Bio-char Solutions, where I could see firsthand the practical applications of biochar production.[76] During my visit, I met with chief operating officer Karl Strahl. Our meeting and tour were eye-opening. As we discussed the biochar production process, Karl explained how the simple transformation of tree waste into biochar captures carbon and generates significant energy. This energy is not just a by-product – it is harnessed, stored and then the energy is sold to California. This is a powerful example of how carbon capture can benefit us both environmentally and economically.

The visit to Oregon Biochar Solutions deepened my understanding of how biochar can play a pivotal role in a sustainable carbon economy. Karl's insights into the energy-generation aspect of biochar production highlighted the full-circle benefits of this process. It is not just about capturing carbon; it is about creating a system where waste is transformed into a resource that powers our homes and businesses, all while contributing to the fight against climate change.

In addition to visiting Oregon Biochar Solutions, I had the chance to meet with Jamie Bartley, the CEO of Unyte Group, a UK-based company leading the way in sustainable and regenerative solutions.[77] Unyte's approach aligns closely with the goals RippleNami has been pursuing, particularly in carbon capture and environmental restoration. Jamie and his team at Unyte are committed to integrating sustainable practices into how we approach environmental challenges. Their work involves taking hemp and tree waste and turning it into biochar and various hemp by-products, creating a full-circle solution that sequesters carbon while producing valuable materials.

Jamie Bartley, CEO of Unyte

Unyte's work is a testament to the idea that sustainability and profitability can go hand in hand, creating systems that benefit both the planet and the economy. Meeting Jamie gave me valuable insights into how these regenerative practices can be implemented on a larger scale. Unyte's dedication to sustainability and innovative approach make it a significant player in combating climate change and promoting environmental stewardship.

With Tafadzwa Dutoit Nyamande, founder of ZimbanjeX

As we venture beyond the traditional biochar giants, let's dive into the adventurous world of ZimbanjeX, a startup in the heart of Zimbabwe.[78] Founded in 2021 by Tafadzwa Dutoit Nyamande, this operation is the most spirited and innovative biochar boutique you might not yet know. And ZimbanjeX doesn't just make biochar – they turn corn cobs and other unexpected heroes like sorghum stalks and hemp stalks into climate-saving superstars with their secret weapon – a former 40-gallon keg kiln designed with a dash of genius and a pinch of moxie.

Corn cob waste being processed by ZimbanjeX

Imagine stepping into ZimbanjeX's world, where every day blends *MacGyver* with *Green Acres*. One minute you're watching a corn cob sceptically eyeing its fiery fate in the retort kiln, and the next you're smack in the middle of a groundbreaking project transforming baobab husks into biochar. Tafadzwa and his team are not just practising science; they are performing alchemy.

The kiln at ZimbanjeX

Joining forces with the Organic Farming Academy and CarbonConnect, ZimbanjeX's projects adhere to what we might call the artisanal toast of carbon sequestration methods, officially known as the Global Artisan C-Sink carbon dioxide removal methodology. ZimbanjeX's vision? To catapult Africa to become a sustainable and pioneering circular economy, one quirky experiment at a time.

With the ZimbanjeX family

So, if you ever find yourself in Zimbabwe, drop by ZimbanjeX. You'll see firsthand how a family-run venture hilariously battles climate change with corn cobs, sorghum and a hearty dose of friendly antics. Don't just plan for a business visit; prepare for an adventure – wear your boots, bring a sense of humour and be ready to dodge flying pigeons. Welcome to the wild side of biochar!

These experiences with Oregon Biochar Solutions, Unyte and ZimbanjeX, and my discussions with industry leaders like Dr Dewar, Tom Miles, Karl Strahl, Jamie Bartley and Tafadzwa Dutoit Nyamande, reinforced the importance of discussing sustainability and actively engaging with the technologies and processes that make it a reality. By visiting these facilities and learning from these experts, I gained invaluable insights into how we can harness natural processes like biochar production to capture carbon, generate energy and contribute to a cleaner, more sustainable world.

CAELUM RESOURCES: A PROJECT DEVELOPER'S PERSISTENCE

Caelum Resources ('caelum' fittingly means 'heavens' in Latin) was founded to create a profound and transparent impact on the carbon credit market.[79] As the chairman and founder, I established Caelum Resources to drive innovation and uplift African lower-income communities through groundbreaking projects centred around industrial hemp cultivation and biochar production.

Based in the US with a keen focus on Africa, Caelum Resources is committed to developing high-impact, low-tech solutions that foster a diversified hemp and biochar industry. Our initiative is built on three core objectives:

- **Industrial hemp economy**: We are spearheading large-scale hemp cultivation to mitigate the adverse effects of tobacco farming and establish a robust infrastructure for producing hemp-derived products, including food, biochar and renewable energy.
- **Certified carbon sequestration**: Leveraging the incredible potential of industrial hemp, which offers carbon sequestration rates 22 times higher than traditional forest-based projects, we aim to provide cost-effective

solutions at just $15 per metric tonne. Our certified biochar production, sourced from agricultural waste, further enhances carbon offsets at $150 per metric tonne.

- **Community improvement**: We are dedicated to improving the living conditions in villages near protected forests by offering essential services such as clean water, internet access and solar power, thereby fostering sustainable development.

IMPLEMENTATION PLAN

The success of Caelum Resources' hemp and biochar carbon credit projects depends on meticulous planning, strategic execution and active stakeholder engagement. We have carefully selected cultivation sites to ensure that our biochar production meets the highest standards of certification and transparency. Through continuous research and innovation, we remain at the forefront of sustainable carbon solutions.

TRANSPARENCY AND IMPACT

At Caelum Resources, transparency and accountability are paramount. We regularly update all stakeholders on project progress, carbon sequestration achievements and socioeconomic benefits. Our projects undergo rigorous environmental and social impact assessments, with comprehensive mitigation measures to ensure positive outcomes. Additionally, we are committed to community development through job creation, skills training and infrastructure enhancements.

LOOKING AHEAD

Having successfully navigated a challenging two-year development phase, as of early 2025 Caelum Resources is poised to enter the critical implementation stage. We are confident in our strategy's potential to deliver significant environmental and economic benefits and remain optimistic about the transformative impact of our innovative approach on both the planet and the communities we serve.

NATURE-BASED CARBON BAMBOO: A DISCUSSION WITH ANETE GAROZA

Nature-based carbon projects are emerging as a transparent path forward in the quest to find sustainable and impactful solutions to climate change. These projects, like the projects involving hemp and biochar, focus on restoring and preserving natural ecosystems, not only removing CO_2 from the atmosphere but also supporting biodiversity, protecting water resources and enhancing the livelihoods of local communities. To delve deeper into this topic, I had the pleasure of speaking with my friend Anete Garoza, a seasoned climate lawyer with a wealth of experience advising large companies on carbon market compliance and ESG (environmental, social and governance) initiatives.

HOW WE MET

My journey to meeting Anete began during a pivotal moment in my work on carbon credits. As discussed in *Chapter 1*, I met with government officials in Zimbabwe to discuss the fallout from the failed Kariba REDD+ project. The project, intended to protect 785,000 hectares of forest, symbolized everything wrong with the carbon credit system – broken promises, unfulfilled benefits for local communities and a general lack of transparency (*Chapter 4* gives more detail). During my meetings, I promised the officials that I would help to educate them on carbon credits and accelerate their participation in the climate economy.

With Anete Garoza, founder of 1MT Nation

This commitment led to the organization of the first Africa Voluntary Carbon Credits Market Forum,[80] held in July 2023 in Victoria Falls, Zimbabwe, where I was invited to speak about the Kariba fraud. Kariba became a case study on transparency – a focal point for much-needed reform in the carbon credit market. During my presentation, as I laid out the uncomfortable truths about the project's failures and the companies that had betrayed Zimbabwe and its communities, a person suddenly appeared on stage and politely asked me to stop talking. This person claimed that I was offending people in the crowd – who, as it turned out, were representatives of the project developer, Carbon Green Investments, and good ole Verra.

You are offending our guest. Could you stop telling the truth about a failed project? Ha, "No." Later at the conference, over the microphone, I said I did not care. I continued my presentation undeterred, and it was clear that the room was filled with both supporters and sceptics.

I met Anete. She hails from Latvia and currently resides in Uganda, where we now often meet and compare notes on our work. Our shared determination to bring integrity to the carbon credit market and our mutual passion for driving real change made us fast friends. From that moment in Zimbabwe onward, Anete and I have worked together to promote transparency and effectiveness in nature-based carbon projects.

Forest coverage in Latvia

Did you know over 53% of Latvia is covered by forests?[81] It is a biodiversity hotspot home to species like lynx and wolves. Anete comes from a family with deep roots in Latvia's forestry sector. Her family's experience spans generations, focusing on sustainable forest management and conservation. The country is a leader in sustainable forestry, with about half of its forests certified for responsible logging. Culturally significant, forests help Latvia act as a carbon sink, absorbing more CO_2 than the nation emits and aiding climate change efforts.[82] Who knew? (You know you love these little side notes!)

1MT NATION:
RESTORING AFRICA'S LANDS AND
FIGHTING CLIMATE CHANGE

Anete co-founded 1MT Nation,[83] which aims to remove 1 million metric tonnes (1 megatonne) of CO_2 from the atmosphere by 2030 through high-quality, nature-based carbon removal projects. Based in Uganda, the organization aims to restore 1 million hectares of degraded land by 2030. 1MT Nation focuses on planting polyculture native and naturalized bamboo on lands that have been without forest cover for over a decade. Polyculture is an agricultural practice where multiple plant species are grown together in the same space. This approach contrasts with monoculture, where only a single crop species is cultivated over a large area. Polyculture mimics natural ecosystems, promoting biodiversity, improving soil health, and reducing the need for chemical fertilizers and pesticides. By collaborating with local landowners, the initiative positively impacts the environment and benefits the communities involved, creating sustainable livelihoods and fostering environmental stewardship.

WHY BAMBOO?

Bamboo is one of the most effective tools in the fight against climate change. Known for its rapid growth and resilience, bamboo is an efficient method of carbon sequestering, producing 35% more oxygen than an average tree.[84] Its ability to capture and store copious amounts of atmospheric CO_2 makes it a powerful ally in reducing greenhouse gas levels globally.

But bamboo's benefits extend far beyond carbon sequestration. Planting bamboo on degraded lands helps to restore ecological balance, conserve and increase biodiversity, and remove CO_2 from the atmosphere. These projects improve the environment and support local communities by creating jobs, offering fair pay and providing additional income for local landowners.

Anete emphasizes the importance of the choice of bamboo for 1MT Nation's projects: "Bamboo's rapid growth and ability to thrive on degraded lands make it an ideal choice for restoration projects. It is not just about planting trees – it is about creating resilient ecosystems that can withstand the challenges of climate change while providing economic opportunities for local communities."

WHY BAMBOO IS WORKING IN UGANDA

Anete Garoza planting a bamboo seedling

In 2023, 1MT Nation successfully began restoring lands in Uganda, marking a significant milestone in its mission to combat climate change and restore degraded lands. Uganda's unique environmental conditions make it particularly well suited to bamboo cultivation. The country's climate supports the rapid growth of bamboo, which quickly establishes itself on degraded lands that would otherwise remain barren. By planting bamboo in these areas, 1MT Nation is sequestering carbon and helping to restore soil health, prevent erosion and create new habitats for wildlife.

The work in Uganda highlights the practical benefits of using bamboo in restoration projects. The species chosen by 1MT Nation are well adapted to local conditions, ensuring that the bamboo thrives and continues providing environmental and economic benefits for years. Moreover, the involvement of local communities in the project ensures that the benefits are shared equitably, with landowners receiving fair pay and additional income through the sale of carbon credits.

CREATING SUSTAINABLE COMMUNITIES

1MT Nation designs its projects to create a sustainable future for local communities. By partnering with local landowners, the company ensures that the benefits of its projects are shared equitably. It creates thousands of jobs through these initiatives while providing steady incomes and improving the quality of life for those involved.

Moreover, 1MT Nation's commitment to sustainability is reflected in its contribution to eight Sustainable Development Goals. These include goals related to poverty alleviation, clean water and sanitation, affordable and clean energy, quality education and more. A sizeable portion of the revenue generated from carbon credits is donated to 1MT Nation's Foundation, which supports local African communities by providing access to clean water, solar energy, primary education, sustainable agricultural practices, and security for crops and villagers.

Bamboo growth at a 1MT Nation project

Anete points out: "We are not just planting bamboo – we are planting the seeds for a better future. Our projects create sustainable livelihoods, improve access to essential services and build resilience against the impacts of climate change."

"Now is the time to act," she urges. "We have laid the groundwork, and the potential for impact is enormous. But we cannot do it alone. We need partners who believe in our mission and are ready to invest in the future of Africa's lands and people."

1MT Nation is more than a carbon removal project developer; it catalyses change, restoring Africa's lands, empowering communities and significantly impacting the global climate. As the company continues to grow, so will its ability to create lasting, positive change. Investing in 1MT Nation is not just about financial returns but about being part of a transformative journey toward a sustainable and equitable future.

THE PATH FORWARD WITH NATURE-BASED SOLUTIONS

Nature-based solutions represent a critical piece of the fight against climate change. As Anete's work with 1MT Nation, Karl's work with Oregon Biochar Solutions, Jamie's work with Unyte and our work with Caelum Resources shows, these projects can deliver real, measurable impacts on carbon sequestration and social benefits. The success of these projects depends on a deep understanding of local ecosystems, close collaboration with communities, and a commitment to creating positive outcomes for both people and the planet. As the carbon credit market continues to evolve, nature-based solutions will play an increasingly crucial role in efforts to mitigate climate change and build a sustainable future.

THE PROVERBIAL FOREST 'ELEPHANT' IN THE ROOM

As discussed in *Chapter 2*, the largest global carbon removal project categories are forest and land use. According to the Berkeley Carbon Trading Project at the University of California, Berkeley, 40% of all carbon credits issued between 2014 and 2024 were related to forest and land use. As depicted in the figure below, there are seven types of forest and land use projects. There are only three project types focusing on Africa: afforestation and reforestation, reducing emissions from deforestation and forest degradation, and sustainable grassland management.

Forest and Land Use

Forest and land use carbon credit types

I want to focus on the second type, often abbreviated to REDD+, which is a framework developed by the UN to combat climate change by incentivizing forest conservation and sustainable management. It concentrates on reducing greenhouse gas emissions by preventing deforestation and forest degradation while promoting the sustainable use of forests and enhancing forest carbon stocks, especially in developing countries. It also emphasizes social and environmental safeguards to ensure benefits reach local communities and contribute to biodiversity. REDD+ forest projects are magnets for fraud, as discussed in *Chapter 4* in relation to the Kariba REDD+ project. What goes wrong? Project developers follow flawed methodologies, and verification companies do not leverage technology to monitor these projects correctly. Since 31% of the world consists of forests, we need to correct the errors of decades past.[85]

I came across some Kenyan nationals who are trying to do just this. Shakil Yusuf and Shemir Yakub started Carbon Credit Africa to fix these wrongs. Based in Mozambique, Carbon Credit Africa is dedicated to combating climate change by curbing illegal deforestation and its detrimental environmental impacts. The company prioritizes the conservation and restoration of forest ecosystems. It aims to develop a sustainable, low-carbon forestry and agricultural economy that benefits some of the poorest communities in developing countries.

The dynamic duo bring rich and inspiring backgrounds to their groundbreaking work. Shakil's deep connection to Mozambique is profoundly personal – his mother is from the country and his grandfather was a celebrated fighter in its struggle for independence. For Shakil, his lineage instils a powerful commitment to the land and its communities, driving him to make a significant impact.

On the other hand, Shemir draws strength from the daring history of the Memon community. His ancestors were among the 200 brave souls who embarked on a perilous journey from India to Kenya in the late 19th century, with only a handful surviving the harsh early days in East Africa. Their story of resilience and entrepreneurship cascades down the generations, shaping Shemir's approach to overcoming today's social and economic challenges. His position as vice president for East Africa of the

World Memon Organization underscores his dedication to community development, seen in his pivotal role in organizing significant events to uplift communities.

Carbon Credit Africa is actively engaged in a carbon credit REDD+ forest project in Mozambique, facilitating the funding, execution and sale of Mozambique carbon credits on international markets. Shakil and Shemir are both leaders and pioneers, steering the project with a vision that marries environmental sustainability with robust community engagement. Their leadership is characterized by a commitment to transparency and empowering local populations, ensuring that the project achieves its environmental goals and fosters a sense of ownership and pride among the local communities. With Shakil and Shemir at the helm, the project is a beacon of innovation and community development, promising a greener future for Mozambique.

With Shakil Yusuf and Shemir Yakub, Carbon Credit Africa, Mozambique

In October 2021, Mozambique distinguished itself as the first country to receive emissions reduction payments from the World Bank–UN Forest Carbon Partnership Facility, acknowledging efforts in the forests within the Gilé and Pebane districts of Zambezia Province.[86] The Zambezia region, rich in mangrove swamps along its 2,500-kilometre Indian Ocean coastline and in dense forests found inland, represents a critical area for conservation efforts.

Zambezia, Mozambique's second most populous province, has an ideal climate for forest conservation projects. Its average annual rainfall is 2,000 mm – among the highest in the country – and its average temperature is 25.6 °C. The province covers an area of 105,008 square kilometres and has a population of approximately 5.1 million.

The REDD+ project in Zambezia includes a variety of activities aimed at enhancing environmental sustainability and community engagement, such as:

- Reforestation for commercial purposes
- Forest restoration
- Development of agroforestry systems and syntropic agriculture
- Distribution of ethanol cooking stoves
- Building of community capacity for sustainable forest management
- Promotion of non-timber forest products
- Support for tourism development
- Intensification of agriculture
- Implementation of solar energy solutions

The project is characterized by a commitment to complete transparency and traceability, with local communities placed at the forefront of the decision-making process.[87] These initiatives underscore Carbon Credit Africa's strategy to mitigate climate change impacts and foster economic growth and sustainability in local communities.

RENEWABLE ENERGY SOLUTIONS: A CONVERSATION WITH DAPHNE DE JONG

Africa faces significant challenges in its power sector, with over 600 million people – almost half the continent's population – lacking access to electricity. Africa's energy deficit hinders economic development, healthcare, education and overall quality of life. The situation is particularly dire in sub-Saharan Africa, where 53% of the population is without electricity. This contributes to extreme poverty and limiting growth opportunities.[88]

To explore the future of renewable energy in the carbon credit market, I turned to my dear friend and expert in the field, Daphne de Jong, founder and CEO of Ven.[89] Daphne and I first met on an expedition to Mount Everest. My goal was to reach base camp and complete the first acclimatization rotation with the climbing team, while Daphne had set her sights on summiting the peak. Despite our different objectives, we hit it off immediately, laughing our way to base camp and beyond. Our friendship is one of a kind, as not many can say they bonded while attempting yoga poses en route to Everest Basecamp.

Daphne's determination, intelligence and sense of humour – especially that sharp Dutch wit – make her a great friend and someone I deeply respect and admire.

Daphne's journey to Everest in 2019 is just one example of her remarkable resilience. Despite her air tank mask malfunctioning, she summited – a situation that would have made most people panic, but Daphne took it in stride. Maybe it's the Dutch stubbornness or, as the Dutch like to say, the ability to keep your wooden shoes firmly planted even in the most challenging situations.

With Daphne de Jong (left), Lobuche Peak East, 6,119 metres, Mount Everest first rotation

Daphne emphasizes that for renewable energy projects to succeed in Africa, they must be designed with local needs in mind, integrating technology with community knowledge, investment with capacity-building, and innovation with sustainability. This approach ensures that these projects are not just top-down initiatives but are aligned with the community's capabilities and aspirations, leading to more sustainable and impactful outcomes.

With Daphne de Jong at Google X

Professionally, Daphne has made an indelible mark in aerospace engineering, starting with an MSc from Cranfield University and another MSc in space technologies from the International Space University. Her career spans contributions to Amazon Prime Air's first customer delivery service in the UK; work with Rivian, SpaceX and Waymo on advanced autonomous vehicles; and recognition by *Forbes* as one of the 30 Under 30 in consumer technology.[90] She is also a former board director at UN Women in San Francisco.

THE IMPORTANCE OF COMMUNITY INVOLVEMENT

In our conversation, Daphne shared her insights on why renewable energy in underdeveloped countries has struggled to take off despite decades of effort. "The promise of renewable energy has been there for a long time, especially in underdeveloped regions," she began. "But the reality is that many of these areas still lack the infrastructure, investment and political stability necessary to implement these solutions effectively." She paused, grinning. "It's like offering someone a windmill without telling them how to get the wind to blow."

She pointed out that while the technology has advanced rapidly, the on-the-ground implementation often needs to overcome significant hurdles. "You need more than just solar panels and wind turbines," she explained. "You need trained personnel, reliable supply chains, maintenance infrastructure and local government support.

Without these, even the best technology can fail to make a lasting impact." Even with these verticals established, traditional resources like liquified natural gas may still be necessary to implement these solutions, and they cannot be quickly phased out. With classic Dutch humour, she added, "It's a bit like trying to make stroopwafels without syrup – they just don't stick together."

Daphne also highlighted the importance of community involvement in renewable energy projects: "Too often, these projects are top-down initiatives where external developers devise a plan but don't fully engage with the local communities. This can lead to projects that don't align with the needs or capacities of the people they're meant to serve, resulting in wasted resources and missed opportunities."

Despite these challenges, Daphne remains optimistic about the future of renewable energy, particularly in the context of the carbon credit market: "There's a growing recognition that for renewable energy to succeed in underdeveloped countries, we need a more integrated approach that combines technology with local knowledge, investment with capacity-building, and innovation with sustainability."

SCALING UP SOLAR AND
WIND PROJECTS

Daphne's insights underscore the complexities of scaling up renewable energy in underdeveloped regions. Solar and wind projects have seen rapid growth globally, but their success in Africa and Asia requires overcoming significant logistical and political barriers. These projects are a natural fit for carbon credits as they directly replace high-emission energy sources with clean alternatives. By generating power from the sun and wind, they prevent millions of tonnes of CO_2 from entering the atmosphere, significantly impacting global carbon reduction efforts.

However, scaling up these projects in underdeveloped regions also means addressing the underlying issues that have hindered their progress for the past 30 years. This involves building local capacity, ensuring reliable supply chains, and fostering strong partnerships with local governments and communities. However, the long-term impact on countries, including their economic and political stability, should be positive for all stakeholders involved.

INNOVATIVE FINANCING MODELS

Innovative financing models are emerging that blend traditional investment with carbon credit revenues to overcome these challenges. For example, project developers can secure funding through green bonds, which are repaid using the proceeds from selling carbon credits generated by renewable energy projects. This approach reduces the financial barriers to entry for renewable energy projects and provides a steady income stream that supports their long-term viability.

Daphne stresses the importance of these new models: "Financing is often the biggest hurdle. By tying renewable energy projects to carbon credits, we create a financial incentive that can attract the necessary investment. But it is not just about the money – it is about ensuring that these investments lead to sustainable, long-term change and that these proceeds are actually used and not sitting there." She quipped, "You know, like finally getting a Dutchman to pay his share – impossible, but once it's done, it's rock solid!"

HYBRID RENEWABLE
ENERGY SYSTEMS

Another innovative solution in the carbon credit market is hybrid systems, which combine solar, wind and storage technologies. These systems maximize energy production by balancing the strengths of different renewable sources, ensuring a more consistent power supply. For example, solar power peaks during the day, while wind power can be harnessed at various times, including at night. Integrating energy storage systems allows the capture and use of excess energy, making these systems more dependable and efficient.

Daphne emphasized that these hybrid systems could be particularly beneficial in regions with variable weather patterns, where reliance on a single energy source may be insufficient. With the use of multiple renewable energy sources, storage will enable flexibility to ensure the required capacity at the correct times. "Hybrid systems offer a way to ensure continuous energy supply, which is crucial for economic development. And in the context of carbon credits, they make the market more resilient and dependable."

MICROGRIDS

Microgrids (decentralized energy grids) offer another exciting opportunity for integrating renewable energy into the carbon credit market. These localized grids can operate independently or with the primary power grid, providing energy security and resilience in areas prone to power outages. Microgrids powered by renewable energy sources (such as solar and wind) can supply clean electricity to remote or underserved communities, reducing their dependence on diesel generators and other fossil fuels.

Microgrids offer a sustainable solution for expanding energy access while combating climate change by generating carbon credits and displacing high-emission energy sources. In addition to offering the most suitable solution for local community independence, microgrids can provide a more secure and resilient option, which is especially crucial during political instability. They also empower communities by giving them greater control over their energy resources, fostering local economic development and improving quality of life.

This is an area where international collaboration can significantly drive progress, particularly concerning the manufacturing and transportation of equipment, and the development of new infrastructure.

NUCLEAR POWER:
A CARBON-FREE OPTION

Nuclear power offers another powerful option for reducing carbon emissions. As one of the most efficient and reliable forms of energy, nuclear power generates large amounts of electricity without producing greenhouse gases. While nuclear energy comes with challenges, such as waste management and high upfront costs, it also provides a stable, carbon-free energy source that can complement renewable energy projects.

Daphne's perspective on nuclear power is particularly relevant. "Nuclear power often gets overlooked in conversations about clean energy," she says. "But it's a critical piece of the puzzle, especially as we look to decarbonize the grid at scale. When combined with renewables, nuclear power can provide the steady, reliable power we need to transition away from fossil fuels completely." While fusion energy is the only option that generates no nuclear waste, we can't afford to wait for these solutions to reach the market. In the meantime, existing options like micro nuclear reactors have proven effective and are ready for deployment. With a laugh, Daphne adds, "It's like having a sturdy Dutch dike to hold back the floodwaters – steady, reliable and built to last."

RENEWABLE ENERGY AND
CARBON CAPTURE INTEGRATION

An emerging trend in the carbon credit market is integrating renewable energy with carbon capture and storage (CCS) technologies. This combination allows for the simultaneous generation of clean energy and capture of CO_2 emissions from industrial processes. For instance, renewable energy can power CCS facilities, reducing the overall carbon footprint of carbon capture operations. Projects that combine renewable energy with CCS offer a powerful tool for achieving net-negative emissions, where more CO_2 is removed from the atmosphere than is emitted.

Admittedly, much more work is needed to improve the efficiency of CCS when bio-based solutions aren't used. Bio-based CCS captures carbon dioxide from bioenergy processes and stores it underground. It can lead to harmful emissions because the biomass absorbs CO_2 as it grows. Capturing the CO_2 when it's converted to energy means removing more CO_2 from the atmosphere than is added, helping to reduce greenhouse gases.

However, making this process cost-effective and efficient is challenging, requiring ongoing research and development to improve and scale up the technology. While CCS is one of the most promising technologies, it is also among the furthest from being economically viable. These projects generate high-value carbon credits by addressing both sides of the carbon equation – reducing emissions at the source and capturing existing atmospheric CO_2.

THE WAY FORWARD WITH
RENEWABLE ENERGY

Renewable energy solutions, including nuclear power, are vital to the carbon credit market and essential for creating a sustainable future. By scaling up renewable energy projects, embracing innovative financing models and integrating advanced technologies, we can significantly enhance the impact of carbon credits while driving global decarbonization efforts. As Daphne highlighted, the success of renewable energy in underdeveloped countries hinges on overcoming the barriers that have held back progress for decades. This means deploying technology and building the infrastructure, training the personnel and fostering the community engagement necessary to make these projects work in the long term. It's not as complicated as we might think. With faith in free markets, better and more affordable solutions will prevail. It is up to the industry and individual governments to ensure that the best solutions become economically viable and accessible to developing countries.

Renewable energy is a beacon of hope for a more sustainable and equitable future. This solution addresses the root causes of climate change while empowering communities and fostering economic growth. Integrating renewable energy and nuclear power into the carbon credit market represents a critical step toward achieving our global climate goals, proving that we can turn the tide in the fight against climate change with innovation.

THE FUTURE OF THE CARBON CREDIT MARKET

The innovations explored in this chapter represent just a few exciting developments in the carbon credit market. As the market continues to evolve, even more creative solutions will emerge – solutions that address the challenges of additionality, permanence, transparency and community involvement while driving real, measurable progress in the fight against climate change.

But while these innovations hold great promise, they require all stakeholders to commit to change. This means embracing modern technologies, adopting new project design and management models, and, most importantly, putting the needs of local communities and the planet at the centre of everything we do.

As we move forward, the carbon credit market has the potential to become a powerful force for good – one that not only reduces emissions but also supports sustainable development, enhances biodiversity and improves the lives of people around the world. But to achieve this potential, we must be willing to innovate, experiment and learn from our mistakes. Only then can we create a carbon credit market that lives up to its promise.

In the next chapter, we will explore the ethical considerations that must guide the future of the carbon credit market. From issues of equity and justice to the responsibilities of corporations and investors, we will look at how we can ensure that the carbon credit market is effective and fair.

CHAPTER 8

ETHICS AND EQUITY IN THE CARBON CREDIT MARKET

As we examine the evolution of the carbon credit market, it is clear that innovation and technology will not save us from the climate crisis. Sure, they are essential, but they are not the whole story. The ethical dimensions – fairness, inclusivity and the actual impact on vulnerable communities – are just as crucial. Without a solid moral foundation, even the fanciest systems will fall flat, leaving us all wondering how we missed the mark.

In this chapter, we will focus on the moral imperatives that must guide the future of the carbon credit market. Creating a functional market is insufficient; we need one that is a beacon of justice and equity. This means protecting and empowering marginalized communities, often the most affected by climate change, through the systems we build. It also means holding corporations and investors to account – no more greenwashing. They need to do more than pay lip service to sustainability; they need to weave it into the fabric of their operations.

We will explore how to construct a carbon credit market that is more than adequate. We need a fundamentally fair market that prioritizes people over profits and contributes to a more just world. This is not simply about ticking boxes to reduce emissions; it is about redefining what it means to build a sustainable future – one where ethics and efficacy are dynamic.

If we get the ethics wrong, we risk perpetuating the injustices we try to eliminate. That would be a serious missed opportunity.

THE ETHICAL IMPERATIVE FOR EQUITY

The carbon credit market holds immense potential to enhance underdeveloped communities by providing financial incentives for sustainable projects benefiting those most affected by climate change. However, with great power comes even greater responsibility, and too many have prioritized profit over people, sustainability and ethics. **This is a call to action – and a call-out – to those who allow greed and complacency to undermine the very purpose of this market.**

Project developers who chase profits at the expense of sustainability perpetuate a system that exploits vulnerable communities, particularly in the Global South. They break promises, exploit lands and mortgage the futures of those who can least afford it, turning what could be a tool for empowerment into another mechanism of oppression.

Stuck in the 1990s, standards governing boards have failed to evolve with the urgency of our climate crisis. Their outdated frameworks allow harmful practices to continue, certifying subpar projects while doing little to ensure accurate, measurable impact. This collaboration in greenwashing undermines trust in the entire system.

Climate funds, intended as beacons of hope and change, often lack the necessary oversight to ensure that the projects they finance deliver on their promises. Without rigorous accountability, these funds risk becoming enablers of exploitation, funnelling money into initiatives that do more harm than good.

Companies that buy worthless carbon credits – out of ignorance or indifference – and keep their mouths shut must be held accountable. These companies claim to be part of the solution while quietly perpetuating the problem. Their silence is deafening and their complicity is shameful. Those who ignore the injustices embedded in the carbon credit market must be called out.

It is not just an oversight; it is a profound ethical failure. The carbon credit market, designed to help mitigate climate change, risks exacerbating the inequalities it was meant to address. The benefits of carbon credits must no longer disproportionately flow to wealthy corporations and investors. The communities on the front lines – the ones contributing the least to climate change but suffering the most from its effects – deserve to be the primary beneficiaries.

Redesigning the market requires correcting these historical imbalances. We must empower local communities, give them control over project development and ensure they receive a fair share of the rewards. We must also demand transparency and accountability from corporations and investors who have long used carbon credits to offset emissions without changing their destructive practices.

The ethical questions surrounding the carbon credit market are not abstract or secondary – they are central to its success or failure. We must challenge the old patterns of exploitation and inequality that have plagued this market for too long. Committing to equity and justice can create a carbon credit market that genuinely aids climate change mitigation and fosters a more just and sustainable world. It is time to stop the exploitation, stop the greenwashing and hold those in power accountable for their actions.

INCLUSION OF MARGINALIZED COMMUNITIES

The carbon credit market is often touted as a win–win: combat climate change while providing economic benefits to local communities. However, the reality on the ground is starkly different, especially for marginalized communities, who are the most directly affected by these projects. These communities often experience significant changes in their land use, resource access and social dynamics due to carbon credit initiatives (as the examples in *Chapters 1–4* show). Yet, despite withstanding the worst of these impacts, they are excluded from the decision-making process and receive only a tiny portion of the benefits.

It is about asking for input and genuinely involving these communities from the get-go. Consultation should be more than a box-ticking exercise; it needs to be a meaningful dialogue that informs every stage of the project, from planning and implementation to management and oversight. The communities should have a real say in how their land and resources are used and how the benefits are distributed.

Moreover, it is crucial to ensure that the benefits – whether from financial compensation, environmental improvements or social gains – are distributed equitably. When done right, carbon credit projects can uplift these communities, providing them with new opportunities and a stake in their development. However, the system fails to deliver on its promises when they are left out or short-changed.

FAIR COMPENSATION AND BENEFIT-SHARING

Let's talk about the money – or, more accurately, follow the rainbow leading to the pot of gold and see who takes it. In too many cases, the profits from carbon credit projects are funnelled into the pockets of a few big players, while the communities hosting these projects are left with crumbs. It is like inviting someone to dinner and giving them the bill while you enjoy it.

Mechanisms must be established to ensure that communities are fairly compensated for their contributions, so as to create a fairer market. Solutions could involve direct payments to community members, investments in local infrastructure (like schools or healthcare facilities) or support for sustainable agricultural practices. Whatever form it takes, compensation should reflect the actual value of what these communities are bringing to the table.

Transparency is key here. Benefit-sharing arrangements should be open and based on clear, mutually agreed criteria. Communities should have a voice in deciding what form of compensation best meets their needs – cash payments, development projects or other forms of support. The days of imposing top-down decisions must end. Instead, communities should be empowered to choose how they benefit from these projects.

PROTECTING LAND RIGHTS
AND SOVEREIGNTY

Land rights and sovereignty are central to the ethical challenges of the carbon credit market. In many parts of the world, land ownership is a complex and contested issue, with traditional and Indigenous land rights often conflicting with formal legal systems. This can lead to disputes over who has the authority to make decisions about land use and who should benefit from carbon credit projects.

To avoid exacerbating these conflicts, it is crucial that carbon credit projects respect and protect the land rights of local communities and Indigenous peoples. All parties should recognize and uphold traditional land tenure systems, ensuring that the community controls how their land is used and preventing land grabs or other forms of exploitation.

In cases where land rights are unclear or disputed, a transparent and inclusive process is essential to resolve the issues before proceeding with a project. This process should involve all relevant stakeholders, including local communities, government authorities and civil society organizations, and should be guided by principles of fairness, equity and respect for human rights.

THE RESPONSIBILITIES OF CORPORATIONS AND INVESTORS

Corporations and investors wield immense power in the carbon credit market. If these two groups genuinely wanted to fix the systemic issues plaguing carbon credits, they could do so; they have the influence and resources to make it happen – swiftly and decisively. Many companies enter this market under the guise of reducing their carbon footprint and showcasing their commitment to corporate social responsibility. However, their actions – or, in many cases, inactions – have profound and often damaging consequences for the very communities and ecosystems they claim to protect. It is time to hold these influential players accountable and demand they use their considerable leverage to drive real, meaningful change rather than perpetuating a system that too often prioritizes profit over people and the planet.

ETHICAL CORPORATE BEHAVIOUR

For corporations, participating in the carbon credit market must be more than just a box to tick or a PR move – it is about taking real responsibility for their environmental footprint and respecting the rights and needs of the communities they impact. Yet, when scandals rocked the carbon credit market in 2023 (as explored in *Chapter 4*), exposing fraudulent projects and exploiting local communities, the silence from companies and investors was deafening. Not a single word of outrage or accountability – just silence.

If corporations cared about their role as stewards of the environment, they would have been the first to demand transparency and integrity in the projects they support. Instead, they prioritized profit over people, paying no heed to the damage caused. Ethical corporate behaviour means more than just compliance with regulations; it is about actively ensuring that investments deliver tangible benefits for the environment and local communities.

Corporations should champion projects that promote sustainable development, enhance biodiversity and genuinely improve the livelihoods of those on the ground. They must avoid involvement in projects that lead to environmental degradation, social displacement or further economic inequality. Ethical behaviour requires complete transparency – being open about the projects chosen, their impacts and the criteria used for their selection. It also means conducting regular, thorough evaluations to ensure investments are making the positive difference they claim to champion.

But again, let's be clear: if these powerful entities wanted to fix the problems in the carbon credit market, they could. They have the resources, influence and reach to demand real change. The fact that they have not speaks volumes.

THE ROLE OF IMPACT INVESTORS

Let's get real – impact investors, who are supposed to be the champions of both financial returns and positive social or environmental outcomes, are a big part of the problem. They strut from one conference to the next, pontificating about their lofty goals and the importance of sustainable investing. But when asked to respond to a simple email or phone call about funding a good carbon credit project? Crickets. If they cannot be bothered to engage with the people on the ground, how can we trust them to drive meaningful change in the carbon credit market?

The real question is: who is giving these impact funds money, and where is the oversight? Impact investors are funnelling capital into projects that are supposed to prioritize equity and sustainability, but what is happening behind the scenes? Are they applying rigorous criteria to ensure these projects genuinely deliver on their promises? The simple response is *no*. Instead, they are just ticking boxes to look good on paper.

It is not enough to throw money at a project and hope for the best. Impact investors need to dig deeper to assess the additionality and permanence of emissions reductions, scrutinize how benefits are shared with local communities, and ensure that the projects they fund contribute to sustainable development in a measurable way. If they are not doing this, they are just perpetuating the same cycle of exploitation and greenwashing that got us into this mess in the first place.

And let's not forget about accountability. These investors have the power to demand greater transparency and integrity in the carbon credit market, but are they using it? They should push for industry standards, advocate for best practices and leverage their influence to reform the system. But they are too busy chasing the next headline or keynote slot to make the fundamental changes that matter.

If these so-called impact investors wanted to be effective, they would be in the trenches, ensuring that every dollar they invest leads to genuine, lasting impact. But as it stands, their silence and inaction are as much a part of the problem as the corrupt systems they claim to be fixing.

THE GLOBAL
EQUITY CHALLENGE

The ethical challenges of the carbon credit market are not confined to individual projects – they also have a global dimension. Climate change is a worldwide problem that affects all of us, but its impacts are not felt equally. The poorest and most vulnerable communities are often those least responsible for greenhouse gas emissions, yet they endure the most climate-related disasters.

This global inequity poses a significant challenge for the carbon credit market. If the market is to be truly ethical, it must address not only the local impacts of carbon credit projects but also the broader issue of climate justice.

ADDRESSING CLIMATE JUSTICE

Climate justice is not just a buzzword – it is a call to action to recognize and address the unequal burdens of climate change. Vulnerable communities, often in the Global South, bear the impacts of climate change despite contributing the least to the problem. According to the UN, the Global South, particularly Africa, South Asia and Central America, is disproportionately affected by climate change. For instance, Africa accounts for less than 4% of global greenhouse gas emissions. Yet, it is one of the regions that is most vulnerable to the impacts of climate change, such as severe droughts, floods and food insecurity.[91] In South Asia, countries like Bangladesh face extreme climate events like cyclones and rising sea levels, threatening millions of lives and livelihoods.

The carbon credit market needs to shift its focus to genuinely promote climate justice. This means channelling financial, technological and other resources into projects that directly benefit communities on the front lines of climate change. Take, for example, the small island nations facing rising sea levels or the rural communities in sub-Saharan Africa grappling with prolonged droughts. These are where carbon credit projects can make the most significant difference, yet they are often overlooked in favour of more profitable or convenient locations.

But it does not stop at the project level. The carbon credit market has the potential – and the responsibility – to drive broader policy changes that address global inequities. For instance, why isn't the market more vocal in advocating for ambitious climate targets that hold significant polluters accountable? Why aren't there stronger pushes for international finance mechanisms that would provide much-needed support for climate adaptation in developing countries? And let's not forget about the global trade and supply chains that continue to pump carbon into the atmosphere at alarming rates

– where is the advocacy for policies that would reduce their carbon intensity?

To put it bluntly, the carbon credit market has the power to promote climate justice, but doing so requires more than just lip service. It requires a commitment to redirect resources to the communities most affected by climate change, push for systemic reforms that address global inequities, and hold corporations and governments accountable for perpetuating these injustices. If those in power wanted to fix the system, they could, and it's high time they started doing so.

THE ROLE OF
INTERNATIONAL COOPERATION

Whether you believe in climate change or not, the reality is that weather patterns are shifting, and we are seeing the effects around the globe. It doesn't matter if you attribute this to human activity or natural cycles – the fact is, the climate economy is here and capital is already in play. So, why not put that capital to work where it is needed most?

Addressing the ethical challenges of the carbon credit market is not about minor adjustments or superficial reforms; it requires bold, decisive action on a global scale. Climate change – or whatever you call it – is a global issue that demands a coordinated response. The carbon credit market can only succeed if it is part of a larger, robust framework of international cooperation that insists on accountability, equity and real impact.

I have attended several climate summits, particularly the COP meetings, and let me tell you – what I have seen feels like a colossal waste of money and effort. COP stands for Conference of the Parties, which is the supreme decision-making body of the United Nations Framework Convention on Climate Change (UNFCCC). The COP brings together countries that have signed on to the UNFCCC, and it meets annually to review progress in combating climate change, negotiate new commitments and discuss emerging issues related to global warming.

The COP meetings are where global climate agreements are discussed and made, including the famous Paris Agreement (during COP21, in 2015). The conferences are attended by leaders, negotiators, scientists, environmental groups and various stakeholders who all come together to figure out how to address climate change.

However, despite the high-level discussions and the intense media coverage these events generate, critics argue that the COP meetings result in more talk than action.[92] The slow pace of meaningful progress has led some to question the effectiveness of these gatherings (which remind me of watching paint dry), especially given the urgency of the climate crisis. The conferences seem more like a show of good intentions than a platform for decisive, impactful action.

These events are like a never-ending talent show – plenty of impressive acts, but nobody wins. I would have been shown the door ages ago if I had run my company like this for 30 years without delivering results. However, the UN and its biggest cheerleaders somehow seem perfectly happy to keep the performance going while the real work sits backstage gathering dust.

International cooperation has to go beyond the usual rhetoric at these climate summits. This means developing and enforcing universal standards and best practices across the carbon credit market – standards that ensure every country, no matter its economic standing, has access to the necessary resources and technologies to participate effectively. No more leaving developing nations behind – it is time for every country to be fully equipped to play a role in this global effort.

But let's cut to the chase: international cooperation is not just about pleasant words and good intentions. It requires absolute transparency and accountability at every level of the carbon credit process. Countries must join forces to expose and eliminate the loopholes, fraud and corruption that have tainted the system for too long. The days of ignoring exploitation must end, and decisive actions need to be taken.

And here is the kicker: without genuine international cooperation, any talk of global equity in the carbon credit market is nothing but a pipe dream. Wealthier nations and corporations must stop hoarding resources and start sharing knowledge and capital. It is time to pool our collective resources and work toward a carbon credit market that reduces emissions in a way that is fair and just for all – no more talk, take action. If the members of the international community genuinely want to fix this system, they have the power to do it. The real question is, do they have the will?

MOVING FORWARD: A VISION FOR AN ETHICAL CARBON CREDIT MARKET

The ethical challenges of the carbon credit market are complex and multifaceted, but they are not insurmountable. By embracing equity, inclusion and justice, we can create a market that reduces emissions and contributes to a more sustainable and equitable world.

The carbon credit market must be guided by a commitment to ethical principles at all levels – from designing and implementing individual projects to developing global standards and policies. This means prioritizing the needs and rights of local communities, ensuring that the benefits of carbon credits are distributed fairly, and working together to address the global inequities that drive climate change.

The innovations and technologies explored in the previous chapters offer exciting possibilities for enhancing the carbon credit market's transparency, effectiveness and fairness. However, these innovations will only be successful if they are grounded in a solid ethical framework that puts people and the planet at the centre of climate action.

As we explore the future of the carbon credit market, let's keep these ethical considerations at the forefront of our thinking. By doing so, we can ensure that the market not only helps to mitigate climate change but also supports the broader goal of creating a more just and sustainable world for all.

The next chapter will examine the practical steps needed to implement these ethical principles in the carbon credit market. From policy recommendations to best practices for project developers and investors, we will explore how to turn our vision for an ethical carbon credit market into reality.

CHAPTER 9

CARBON CREDIT CREDIT SHAKE-UP: BUILDING A SYSTEM THAT WORKS

The carbon credit market is at a critical crossroads, and we can no longer afford to play by the old rules. Traditional approaches to managing carbon credits have proven to be deeply flawed and riddled with inefficiencies, inequities and outright fraud. It is time for a radical overhaul – a reimagining of the carbon credit system that does not just patch up the cracks but builds something new from the ground up.

RETHINKING NATIONAL CARBON FRAMEWORKS: TAKE CONTROL AND ENFORCE ACCOUNTABILITY

Let's start with a bold idea: nationalizing the carbon credit framework. Yes, you heard that right. Each country, particularly those in the Global South, should take complete control of its carbon credit market. They should establish and enforce strict national frameworks that govern how carbon credits are generated, traded and used. This is not about nationalism but sovereignty and ensuring that the benefits of carbon credits stay where they belong – within the countries that generate them.

The majority of the Global South nations do not have the frameworks to effectively manage and regulate carbon credits. This is unacceptable. Every country must have a national carbon strategy that clearly outlines who controls the credits, how projects are monitored and what happens when things go wrong.

Here is the kicker: if a foreign company wants to come in and develop a project, they must partner with local entities and adhere to the strict national guidelines. No more unchecked access, no more broken promises.

And if they fail to deliver on their commitments, they should face harsh penalties – think heavy fines, asset seizures or outright bans from operating in the country. This is not about being punitive; it is about ensuring that carbon credits are a force for good rather than just another avenue for exploitation.

REDEFINING INTERNATIONAL COOPERATION: ENOUGH TALK, MORE ACTION

International cooperation has been the buzzword at every climate summit, but let's be honest – most of it has been more about photo ops than actual progress. We have had enough of grand speeches and endless reports; we need action now.

The carbon credit market must be embedded within a robust international cooperation framework that goes beyond setting standards. We need actual enforcement mechanisms. Countries should work together to create an international watchdog with teeth – an entity that can impose sanctions on companies and nations that violate the principles of transparency, equity and sustainability.

And let's not shy away from the tough conversations: whether you believe in climate change or not, weather patterns are changing and the global economy is shifting toward climate action. Again, this is not about ideology; it is about pragmatism. The carbon credit market is growing, so let's put the existing capital to work where it is needed most rather than allowing it to swirl around in perpetual unproductive loops.

EMPOWERING LOCAL COMMUNITIES: FROM PASSIVE BENEFICIARIES TO ACTIVE STAKEHOLDERS

The traditional carbon credit market treats local communities as passive beneficiaries – recipients of whatever crumbs are left after the big players take their share. It is time to flip the script. Communities should be the driving force behind carbon credit projects, not just participants but leaders.

Imagine a carbon credit market where communities own and manage the projects themselves. This is not just dreaming; it is entirely possible. Governments and NGOs should provide the initial capital, training and resources but step back and let the communities take over. These projects should be tailored to the specific needs of the communities, whether that's reforestation, sustainable agriculture or renewable energy.

Let's make sure they are getting paid properly. Benefit-sharing should not be an afterthought; it should be the first thing on the table. Communities should receive direct financial compensation, and they should have control over how that money is spent. Want to build a school?

Great. Need clean water systems? Let's do it. The key is to empower communities to make these decisions themselves rather than having them dictated from the outside.

OUT WITH THE OLD, IN WITH THE NEW: RADICAL TRANSPARENCY AND INNOVATION

The old methods for monitoring and verifying carbon credits are slow, cumbersome and easily manipulated. Radical transparency and innovative technology are needed to bring the carbon credit market into the 21st century.

Let's talk about blockchain – not the buzzword, but the technology. Imagine a world where every carbon credit is tracked on a transparent, immutable ledger: no more shady deals, double counting or fraud. Every transaction is visible to everyone, and any attempt to cheat the system is immediately flagged and addressed.

But that is just the beginning. We should use AI and machine learning to predict and optimize carbon sequestration, satellite imagery to monitor projects in real time, and even drones to ensure that what is happening on the ground matches what is being reported. Technology exists; it is time to use it.

And let's not forget about the importance of simplifying processes. The current carbon credit market is a maze of bureaucracy, with endless forms, audits and reports that often obscure the truth more than reveal it. We need a streamlined system that is easy to understand, easy to use and impossible to game.

REVOLUTIONIZING THE ROLE OF INVESTORS: PUT YOUR MONEY WHERE YOUR MOUTH IS

Investors, it is time to step up. You've been talking a big game about sustainability and impact, but now it is time to back that up with real action. The days of mindlessly throwing money at any project with a green label are over. To be effective, you must be strategic, rigorous and ethical.

Start by conducting due diligence. Look beyond the glossy brochures and slick presentations – dig deep into the projects you fund. Are they delivering the emissions reductions they promise? Are the communities benefiting or are they being exploited? If the answers are not satisfactory, walk away.

And don't just invest – advocate. Use your power and influence to push for greater transparency and accountability in the carbon credit market. Support initiatives that promote ethical practices, and don't be afraid to call out immoral behaviour when you see it. If investors demanded better, the market would change – period.

MOVING BEYOND LIP SERVICE: EDUCATION, COLLABORATION AND CAPACITY-BUILDING

It is time to stop talking about education and capacity-building as nice-to-haves and start treating them as essentials. Communities cannot participate in the carbon credit market if they do not understand how it works. They need training, resources and ongoing support, not just a one-off workshop.

However, education is not just for the communities. Governments, NGOs and even corporations need to be educated about what a just and equitable carbon credit market looks like. This means breaking down the silos, sharing knowledge and working together to build a system that functions for everyone.

Here is a radical idea: let's create an international online carbon credit academy where stakeholders worldwide can learn, share and innovate. Imagine a place where the best minds in environmental science, economics, technology and social justice come together to create the future of the carbon credit market.

A BLUEPRINT FOR ETHICAL ACTION

The steps outlined in this chapter are not just about improving the carbon credit market – they are about transforming it into something that works. By embracing radical transparency, empowering communities and holding everyone accountable – from governments to investors to project developers – we can create a market that reduces emissions and builds a more just and sustainable world.

As we move forward, we must remain committed to these. This is not just about tweaking the system but about reimagining it entirely.

In the next chapter, we will explore the future of the carbon credit market and how these bold ideas will shape its evolution. We will also look at emerging trends and the potential for innovations to enhance the market's effectiveness and impact. The future is uncertain, but one thing is clear: the carbon credit market must change, and we must be the ones to change it.

CHAPTER 10

30 YEARS OF LESSONS LEARNED AND INNOVATIVE STRATEGIES FOR THE FUTURE

With Stanley Mathuram (left) and Obert Dube (Zimbabwean Poet)

Remember Stanley Mathuram, the fascinating individual I mentioned back in *Chapter 1* who sent me the article that started my journey toward finding out about these crappy carbon credits? Well, he is back, and this time, he is not just a character in our story – he is the guide we need as we navigate the lessons learned over the past 30 years and strategize for the future of the carbon credit market.

Stanley is not just a bystander in the sustainable industry. He holds a BS in chemical engineering from the

National Institute of Technology, Tiruchirappalli, and an MS in environmental engineering from Michigan State University, another rivalry of my beloved University of Michigan football team (again, I say Go Blue!). But Stanley is not just all brains – he is six-foot-three tall, a former Ford model and a former tennis pro. Honestly, he is the kind of guy who makes you question whether he was secretly created in a lab. He is an absolute walking anomaly. So, when I say he is a fascinating individual, I mean it. Stanley brings over 25 years of experience in sustainability, business development, circularity, climate finance, carbon markets, the transition economy and environmental engineering. His technical expertise and relentless pursuit of real impact have made him a leading voice in the field.

EARLY DAYS: BIG IDEAS, BIGGER DISAPPOINTMENTS

Stanley recalls the early days of carbon credits with a mix of nostalgia and frustration. "Back then," he says, "we were all about the superb ideas. Save the planet? Sure, let's do it. But we got bogged down in bureaucracy somewhere along the way and missed the bigger picture."

His insights are invaluable as we look back at the industry's evolution – or lack thereof. He has witnessed the carbon credit market's transformation from a promising concept into a multi-billion-dollar industry. Yet, he also sees how that transformation has not always translated into real-world impact. "We started with noble intentions, but the execution often left much to be desired," Stanley notes.

Over the past three decades, the global community has undertaken numerous initiatives to combat climate change, with varying degrees of success. Unfortunately, some of the most significant setbacks have been due to systemic flaws and fraudulent practices within the carbon credit market – many of which were perpetuated by large climate companies and outdated standards governing bodies. These entities, often mired in bureaucracy and resistant to technological innovation, have contributed to a market that is frequently more focused on maintaining the status quo than driving real, impactful change.

Stanley's frustration with the stagnation in the industry has led him to take the reins at ECG Global Solutions, where he is bringing his vision of a more effective and ethical carbon credit market to life. "I have had enough of the same old song and dance," he says. "At ECG, we are doing things differently. We are leveraging technology, transparency, and a commitment to real impact to create a working carbon credit market."

WHAT HAS BEEN DONE: KEY INITIATIVES OF THE PAST 30 YEARS

Let's examine what has been done over the past 30 years, why it has not worked, and how the entrenched systems and consultants within the carbon market need to be overhauled.

THE ROLE OF STANDARDS GOVERNING BODIES

Standards governing bodies, such as the Gold Standard and the Verified Carbon Standard, were established to ensure that carbon credits are credible and verifiable, and contribute to real emissions reductions. However, these bodies have become increasingly archaic over time, hampered by bureaucratic inertia and a lack of innovation.

Stanley puts it bluntly: "These organizations rely heavily on a sprawling network of consultants, many of whom are more invested in maintaining their relevance and income streams than driving meaningful change. As a result, the processes for verifying and certifying carbon credits have become convoluted, slow and often disconnected from the realities on the ground." This environment has become ripe for fraud and manipulation by those more interested in profit than genuine environmental impact.

Why did it not work? The standards governing bodies have failed to keep pace with technological advancements and evolving market needs. Their overreliance on analogue outdated methods and the entrenched interests of consultants have stifled innovation and allowed systemic flaws to persist, undermining the credibility and effectiveness of the carbon credit market.

THE CLEAN DEVELOPMENT
MECHANISM

The Clean Development Mechanism (CDM), established under the Kyoto Protocol,[93] was intended to help developed countries meet their emissions targets by financing projects that reduce emissions in developing countries. However, CDM quickly became emblematic of the carbon credit market's problems.

"Under the CDM, projects are often approved based on outdated methodologies and questionable assumptions," Stanley says, shaking his head. "The standards governing bodies, heavily influenced by consultants who had little incentive to innovate, allowed many projects to proceed despite glaring issues such as inflated emissions reductions and lack of additionality. As a result, the CDM became synonymous with greenwashing and fraud."

Why did it not work? The CDM was hampered by over-reliance on consultants and rigid standards that failed to adapt to the latest information or technological advancements. This allowed large climate companies to manipulate the system, securing approvals for projects that did little to reduce emissions while lining their pockets with profits. Are you seeing a trend yet?

THE EUROPEAN UNION EMISSIONS TRADING SYSTEM

The European Union Emissions Trading System (EU ETS)[94] is the world's largest carbon market and was designed to cap emissions and allow companies to trade allowances. However, like the CDM, it has been plagued by issues related to overallocation, low prices and fraud.

"The standards governing bodies overseeing the EU ETS were slow to adapt to market changes, allowing large corporations to exploit loopholes and engage in practices such as VAT fraud and the resale of fake carbon credits," Stanley explains. "The involvement of consultants, who were more focused on navigating the system than improving it, further compounded these issues."

Why did it not work? The EU ETS's reliance on outdated standards and a complex web of consultants led to a system that could easily be exploited by those with the resources to do so. The lack of flexibility and innovation in the market's governing bodies allowed these problems to persist, undermining the system's effectiveness.

THE PARIS AGREEMENT AND
THE STANDARDS BOTTLENECK

The Paris Agreement shifted toward nationally determined contributions[95] and a more decentralized approach to emissions reductions. However, the standards governing bodies, and their army of consultants, have struggled to adapt to this new reality.

Stanley points out: "Instead of embracing innovative technologies and methodologies that could streamline the verification process and ensure the integrity of carbon credits, these bodies have clung to outdated, crusty practices. This has created a bottleneck in the system, slowing the approval of new projects and stifling innovation."

Why did it not work? The failure of standards governing bodies to evolve has created significant barriers to successfully implementing the Paris Agreement. The continued reliance on a cadre of consultants more interested in maintaining the status quo than embracing the latest ideas has further hampered progress.

CALLING OUT THE ARCHAIC STANDARDS GOVERNING BODIES AND CONSULTANTS

The carbon credit market's problems are not just technical but structural. With rigid procedures and outdated thinking, the standards governing bodies have become a significant obstacle to progress. The consultants who have built their careers within these systems are often more interested in preserving their relevance than fostering meaningful change. It is time to address these issues and advocate for a complete system overhaul.

ARCHAIC STANDARDS GOVERNING BODIES

Stanley is particularly vocal on this topic: "The standards governing bodies have become relics of a bygone era. Their failure to embrace modern technologies, such as digital monitoring, AI-driven verification and blockchain for traceability, has kept the carbon credit market mired in inefficiency and fraud. These bodies must adapt to the modern era or be replaced by more dynamic organizations capable of driving the change needed to meet global climate goals."

CONSULTANT-DRIVEN CULTURE

The carbon credit market has been heavily influenced by a network of consultants with little incentive to innovate. Stanley's disdain for this consultant-driven culture is palpable: "These consultants often perpetuate outdated practices, providing services more about navigating the existing system than improving it. This consultant-driven culture has created a stagnant market resistant to change."

Now, the crux of the issue: the market needs to break free from this consultant-driven mindset to move forward. This can be achieved by streamlining processes, reducing reliance on intermediaries, and embracing technologies that enable more direct, transparent interactions between project developers, buyers and regulators.

FAILURE TO EMBRACE INNOVATION

One of the most glaring failures of the past 30 years has been the refusal of standards governing bodies and their consultants to embrace innovation. "Whether it is the use of advanced data analytics for project monitoring, AI for real-time verification or blockchain for transparent credit tracking," Stanley says, "the tools exist to make the carbon credit market more efficient and credible. Yet, these innovations have been ignored to maintain the status quo."

CALL FOR A
NEW APPROACH

Stanley's blueprint for the future is built on the lessons learned from the past 30 years. He believes that by focusing on transparency, accountability and impact, we can make a carbon credit market that reduces emissions and contributes to a more just and sustainable world.

The carbon credit market needs a novel approach prioritizing transparency, efficiency and genuine environmental impact over bureaucracy and profit. This means dismantling the current system of outdated standards and consultant-driven practices and replacing it with a more dynamic, technology-driven framework.

"Digital monitoring, reporting and verification (MRV) systems should be at the forefront of this new approach," Stanley says. "These systems can provide real-time data on project performance, ensuring that credits are issued based on actual emissions reductions, not on manipulated or outdated estimates."

INNOVATIVE IDEAS AND STRATEGIES FOR THE FUTURE

Having identified the structural flaws that have hampered the carbon credit market, it is essential to propose new strategies to make it more effective and equitable. The following ideas represent a bold departure from the past and offer a way forward that embraces innovation and integrity.

DIGITAL MONITORING, REPORTING AND VERIFICATION SYSTEMS

To replace the outdated practices of standards governing bodies, we must implement digital MRV systems that use satellite data, AI and blockchain to provide real-time, transparent tracking of carbon credit projects. These systems eliminate the need for costly, slow-moving consultants and ensure carbon credits are based on verifiable, up-to-date information.

INDEPENDENT OVERSIGHT AND DECENTRALIZED VERIFICATION

Instead of relying on a centralized standards body, the carbon credit market should embrace decentralized verification systems that leverage a network of independent auditors, local communities and technology platforms. Stanley advocates for this approach, as it would reduce the risk of fraud and increase the credibility of carbon credits.

A NEW ROLE FOR TECHNOLOGY-DRIVEN COMPANIES

Companies like Fiùtur, with its SMART system (see *Chapter 5*), and RippleNami, with its Carbon Clarity system (see *Chapter 6*), offer a model for how technology can streamline and enhance the carbon credit market. By integrating data from multiple sources and providing a transparent platform for managing carbon credits, these companies can replace the outdated standards governing bodies with a more effective system.

ELIMINATING THE
CONSULTANT-DRIVEN CULTURE

The carbon credit market must avoid dependence on consultants who profit from complexity and inefficiency. Stanley's vision is clear: "By simplifying processes, reducing administrative burdens and embracing technology, the market can become more accessible and efficient, allowing greater participation from diverse stakeholders."

EMPOWERING LOCAL
COMMUNITIES

Future carbon credit projects should be designed with the active participation of local communities, ensuring that they benefit from them and have a say in how they are managed. Stanley insists that shifting from top-down, consultant-driven approaches will lead to more sustainable and equitable outcomes.

BUILDING A MARKET FIT FOR THE FUTURE

The past 30 years have shown us what doesn't work in the carbon credit market: outdated standards, consultant-driven practices and a failure to embrace innovation. To build a market that is fit for the future, we must be willing to dismantle these structures and replace them with systems that prioritize transparency, accountability and genuine environmental impact.

By adopting modern technologies, empowering local communities and eliminating unnecessary intermediaries, we can create a carbon credit market that reduces emissions and supports sustainable development and social equity. It is time to leave behind the old ways and embrace a new vision for the carbon credit market – one that is truly capable of addressing the challenges of climate change.

In the concluding chapter, we will reflect on our journey throughout this book and consider the role that each of us can play in shaping the future of the carbon credit market. We are ready to lead the charge into this new era,

and it is now clear that we need more people like Stanley
– those who are not afraid to challenge the status quo and
push for the change we desperately need.

CHAPTER 11

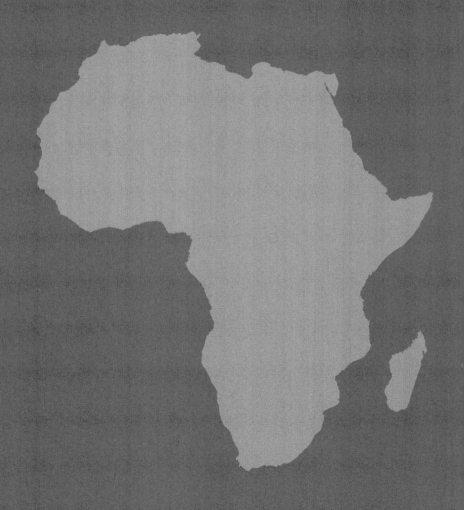

REFLECTIONS AND THE ROAD AHEAD

As we close this exploration of the carbon credit market, it is crucial to reflect on the lessons learned and consider the road ahead. This journey began with the promise of carbon credits – a concept that seemed like a silver bullet in the fight against climate change. Yet, as we have uncovered throughout this book, the reality is far more complex. The system, while well intentioned, has been plagued by inefficiencies, corruption and a failure to deliver on its promises to those most in need.

THE PROMISE
AND THE PERIL OF
CARBON CREDITS

In the first chapter, we delved into the bold promise of carbon credits and the harsh reality of greenwashing that undermines the entire system. Carbon credits were supposed to be a straightforward solution: companies could offset their emissions by investing in projects that reduce or absorb CO_2 elsewhere, often in developing regions like Africa. The concept is as appealing as it is simple – continue business as usual but with a clean conscience.

However, as we journeyed through the stories of communities like those led by Chief Mola, we saw firsthand the gap between promise and reality. The €30 million that was supposed to transform his community never materialized meaningfully. Instead, the local people, who should have been the primary beneficiaries, were left with little more than empty promises. This disillusionment is not isolated; it is a widespread problem that points to a deeper systemic issue.

UNMASKING THE GATEKEEPERS OF THE CARBON CREDIT MARKET

As we navigated the middle chapters, it became clear that the institutions designed to protect and regulate the carbon credit market often need to do more. Standards governing bodies, project developers and climate investors are supposed to be the gatekeepers, ensuring that carbon credits are credible and that the associated projects deliver real, measurable benefits. Yet, as we discovered, these entities are often more concerned with preserving their status and profits than driving meaningful environmental and social change.

These gatekeepers have created a system that, in many cases, benefits everyone except those it was designed to help. The rigorous standards and meticulous monitoring they boast about are, in many instances, little more than a facade – an illusion of accountability that hides the true extent of the market's failings. The problem is not just technical but also profoundly moral, as those who should benefit most are often left with the least.

THE ROLE OF UNSUNG HEROES IN CONSERVATION

Despite the systemic failures, we also encountered stories of hope and resilience. Individuals like Steve Edwards and Damien Mander are unsung conservation heroes, working tirelessly to protect wildlife and support local communities with minimal recognition or support from the carbon credit market. These are the people who are making a tangible difference on the ground. Yet, their efforts are often co-opted by project developers who take credit for their work without providing the necessary financial or logistical support.

These stories underscore the importance of genuine, grassroots-driven conservation efforts. They remind us that real change comes not from top-down initiatives but from those who live and breathe the land they are working to protect. If the carbon credit market is to succeed, it must do more to support these local heroes and ensure that they are not just a footnote in the global climate narrative but central to it.

UNCOVERING THE CARBON CREDIT BETRAYAL

As related back in *Chapter 1*, my meeting with government officials revealed deep outrage and betrayal over the carbon credit projects. Officials were shocked to learn that foreign companies had made over €100 million from their trees without delivering the promised €30 million to their communities. Both the government and local communities felt deceived, as they had been led to believe that carbon credits would bring significant benefits relating to education, healthcare and infrastructure. Instead they were left empty-handed, which highlights a systemic failure by the climate companies involved. This betrayal has eroded trust and will complicate future efforts to implement effective climate action.

THE NEED FOR A NEW APPROACH

The carbon credit market needs a fundamental overhaul as we look to the future. The current system is unsuitable because it relies on outdated standards and consultant-driven practices. We need a novel approach that prioritizes transparency, accountability, and genuine impact over bureaucracy and profit.

One key takeaway from this book is the need to embrace innovation. Technologies like AI, satellite monitoring and blockchain can revolutionize how we track and verify carbon credits, ensuring they are based on actual, measurable emissions reductions.

I purposely did not dive into blockchain in this book because it is a valuable tool for specific applications but has notable limitations, such as scalability, energy consumption and regulatory concerns. Blockchain networks, particularly those using energy-hungry mechanisms, can consume around 130 terawatt-hours of electricity yearly. That's roughly the same amount of energy a country like Argentina uses in a year, or enough to power millions of homes. In terms of carbon emissions, a blockchain network generates about 40 to 50 megatonnes of CO_2 annually. That is comparable to the emissions of a country like New Zealand. It would require around 50 million carbon credits to offset this, equivalent to planting around 800 million trees.[96]

Blockchain is an OK solution for now, but alternative technologies are being developed with reduced energy consumption and environmental footprints. Examples include proof-of-stake networks, directed acyclic graphs, quantum computing and more advanced distributed ledger systems. These are already under development and promise to surpass blockchain in terms of efficiency, speed and scalability. These innovations will eventually replace blockchain as the go-to solution.

Digital monitoring, reporting and verification systems, championed by companies like Fiùtur, represent the future of the carbon credit market. These systems can provide the transparency and accountability that has been sorely lacking, making it easier to ensure that the benefits of carbon credits reach those who need them most.

But technology alone is not enough. We also need a shift in how we think about carbon credits and their role in the global fight against climate change. This means moving away from a top-down, profit-driven model and toward a more inclusive and community-centred one. Local communities must be at the heart of carbon credit projects, not just as passive beneficiaries but as active participants who have a say in how these projects are designed, implemented and managed.

THE ROLE OF INTERNATIONAL COOPERATION

Achieving this vision will require unprecedented international cooperation. Climate change is a global problem that demands a coordinated response, and the carbon credit market can only be effective if it is embedded within a broader framework of international climate action. Governments, international bodies, corporations, investors and NGOs must all work together to develop common standards, share knowledge and ensure that the benefits of carbon credits are distributed equitably.

However, as discussed in *Chapters 8* and *9*, the current state of international cooperation leaves much to be desired. Events like the annual COP (Conference of the Parties) climate summits are filled with high-minded speeches and grand promises, but the results often fall short. The UN and its biggest fans seem content to meander along, making little progress while the world keeps turning. Whether you believe in climate change or not, there is a climate economy, and the weather has changed in some areas. It is time to put the existing capital to work where there is the greatest need.

International cooperation must go beyond talking points. We must develop and enforce common standards and best practices across the carbon credit market – standards that ensure every country, regardless of economic standing, has access to the resources and technologies necessary to participate effectively. No more leaving developing nations behind; every country must be equipped to play a role in this global fight.

Moreover, genuine international cooperation involves more than just handshakes and photo ops. It demands genuine transparency and accountability at every level of the carbon credit process. Countries must work together to expose and eliminate the loopholes, fraud and corruption that plague the system. The days of paying no heed to exploitation must end.

Let's be clear: without international cooperation, the dream of global equity in the carbon credit market is dead on arrival. Wealthier nations and corporations must stop hoarding resources and start sharing knowledge and capital. It is time to pool our collective resources and work toward a carbon credit market that reduces emissions and does so in a fair and just way for all – no more lip service, just action.

AFTERWORD

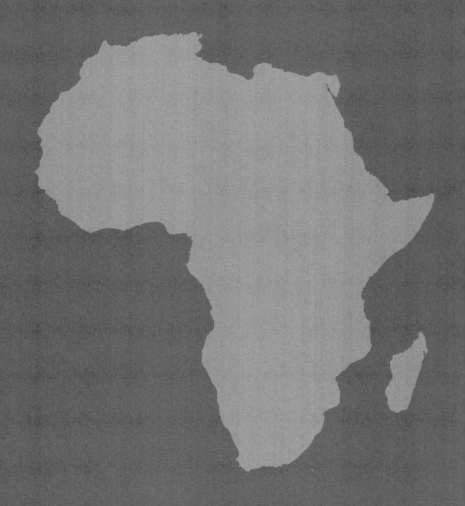

THE
URGENCY
IS REAL

As we close this journey through the tangled mess of carbon credits, let me clarify again: I don't care whether you agree or not if climate change is real or not. What matters is that a $2 trillion climate economy is shaping our future, and the people protecting our planet – especially in underdeveloped communities – deserve to benefit from it. These communities have been exploited for too long, left with empty promises while corporations and consultants line their pockets. It's time to end this cycle.

The carbon credit market's future is filled with challenges and opportunities. This market has the potential to be a powerful tool in the fight against climate change, but only if it operates transparently, accountably and equitably. To achieve this, we must dismantle outdated structures holding the market back and embrace innovative approaches prioritizing real impact over profit.

The journey ahead will not be easy, but it is a path we have to walk together. Whether we are government officials, corporate leaders, investors, project developers or concerned citizens, we all have a role in shaping the future of the carbon credit market. By working together, we can create a system that reduces emissions, supports sustainable development and protects vulnerable communities. The communities I have worked with, especially in Africa, face severe challenges that require everyone – countries, companies and people alike – to unite for a greener, fairer world.

Navigating carbon credits is complex, but my mission is clear: to educate, drive awareness and build a better system. I focus on Chief Mola and communities like his – those who were promised much and given little. Let's turn these crappy carbon credits into something meaningful, something that genuinely uplifts these communities.

BE PART OF
THE CHANGE

We are standing at the crossroads of a pivotal moment in history, where the choices we make today will determine our planet's future and the course of countless people's lives worldwide. The carbon credit market has the potential to be a powerful force for good, but only if we embrace transparency, accountability and ethical practices. It is up to all of us – governments, businesses, investors, project developers and concerned citizens – to ensure this system delivers on its promises.

My challenge to you is to get involved. Contact me if you are uncertain of how to proceed via my website **www.jayeconnolly.com**, and we'll figure it out together.

Here's how you can get involved:

- **Do your homework**: If you're buying carbon credits or supporting sustainability initiatives, know exactly where your money goes. Treat carbon credits like any other investment – demand traceability and transparency. Know where each carbon credit originated from and what impact it has. Hold companies and project developers accountable for delivering real, measurable results.

- **Support ethical projects**: Whether you're an investor, a company or an individual, choose to support carbon credit projects that prioritize the needs of local communities and deliver tangible benefits. Seek out community-driven initiatives, employ sustainable practices and contribute to reducing emissions.

- **Push for change**: It's time to eliminate the unnecessary layers of consultants between project developers and certified credits. With technology at our fingertips – such as AI, blockchain and real-time monitoring – there is no reason these middlemen should continue to siphon off resources and slow down progress. Advocate for technology-driven systems that can directly connect projects to certification bodies, ensuring quicker, more efficient and more transparent processes. This shift will free up resources to support the real work on the ground – projects that reduce emissions and benefit communities.

- **Amplify local voices**: Ensure that the voices of marginalized and underdeveloped communities are heard. Advocate for their inclusion in decision-making processes and demand fair compensation and benefit-sharing in carbon credit projects. These communities are our planet's front-line protectors – let's ensure they are empowered and supported.

- **Educate yourself and others**: Learn more about how carbon credits work, the challenges in the market and the innovative solutions emerging to create a more equitable system. Share this knowledge with others through conversations, social media or community action. Education is the foundation of real change.

- **Reach out**: If you're unsure about a project, a company or how to get involved, reach out. This journey is complex, but you don't have to navigate it alone. Whether you're questioning a carbon credit's legitimacy or want to learn more about meaningful climate solutions, connect with others committed to driving positive change.

- **Be a responsible global citizen**: Remember, it's not just about reducing emissions – it's about building a just, equitable and sustainable future for everyone. Every action you take can contribute to this vision, whether in your personal life, your community or business. Take responsibility for your role in the global effort to combat climate change and promote sustainable development.

This is your moment to step up and be part of the solution. The carbon credit market – and the broader climate economy – will only succeed if it works for everyone, especially the underdeveloped communities on the front lines of climate protection. It's time to hold the system accountable, demand real impact and build a fairer, more transparent future.

So, the next time someone asks you to support a carbon credit initiative, think critically, ask questions and act with intention. Let's turn these so-called crappy carbon credits into a golden opportunity for all.

Join me in this fight, and let's make it right.

Thank you.

GLOSSARY OF TERMS

Additionality

A key concept in carbon offsetting, referring to whether a carbon reduction project would have happened without the financial support generated by the sale of carbon credits. If the emissions reduction would have occurred anyway, the project should not be considered 'additional' and should therefore not be eligible for carbon credits.

Afforestation

Planting trees on land that has yet to be forested. This is often done to create a carbon sink. Trees absorb CO_2 from the atmosphere, thus mitigating climate change.

Carbon capture and storage

A technology used to capture CO_2 emissions from industrial processes or the atmosphere and store them underground in geological formations to prevent them from being released back into the atmosphere.

Carbon credits

A tradable certificate or permit representing the right to emit one metric tonne of CO_2 or the equivalent amount of another greenhouse gas. Carbon credits are generated by projects that reduce, capture or avoid greenhouse gas emissions.

Carbon footprint

The total amount of greenhouse gases, primarily CO_2, emitted directly or indirectly by an individual, organization or product over a defined period.

Carbon market

A market where carbon credits are bought and sold, often to comply with emissions reduction commitments or to voluntarily offset carbon footprints. The carbon market includes compliance markets (e.g. the European Union Emissions Trading System) and voluntary markets.

Carbon offset

Reduction of greenhouse gas emissions, often through renewable energy, reforestation or conservation projects, to compensate for emissions produced elsewhere. Carbon credits frequently represent carbon offsets.

Carbon sequestration

The process of capturing and storing atmospheric CO_2, either naturally (e.g. through forests) or through technological solutions like biochar or carbon capture and storage, to reduce the amount of greenhouse gases in the atmosphere.

Clean Development Mechanism (CDM)
A framework established under the Kyoto Protocol to allow industrialized countries to invest in emissions reduction projects in developing countries to meet their emissions reduction targets.

Compliance market
A market where countries and companies purchase carbon credits to offset their emissions in accordance with their legal obligations, as opposed to the voluntary carbon market, where purchases are entirely optional.

Consultant-driven culture
A term used for a system or industry, like the carbon credit market, where consultants and intermediaries have significant control over the processes and standards, often prioritizing complexity and profits over effectiveness and transparency.

Decentralized verification
A verification process that leverages independent auditors, local communities and technology platforms to assess carbon credit projects. This is an alternative to traditional centralized bodies, which may be slower and less transparent.

Deforestation
Permanent removal of forests to make land available for other uses, such as agriculture or urban development. It is a significant contributor to global CO_2 emissions.

Directed acyclic graph (DAG)

A structure where transactions are linked directly to each other, unlike traditional blockchains with blocks. This allows transactions to be processed simultaneously, making transaction procession faster and more scalable. DAGs are also more energy-efficient since they don't require heavy mining processes.

Emissions reduction

The process of reducing emissions from polluting sources (e.g. fossil fuels) or enhancing processes that absorb carbon (e.g. planting forests). Such measures can decrease the amount of greenhouse gases released into the atmosphere.

European Union Emissions Trading System (EU ETS)

The most extensive carbon trading system in the world, intended to reduce greenhouse gas emissions. It operates as a cap-and-trade system, where companies must hold permits to cover their emissions and those that reduce their emissions can sell excess permits.

Greenwashing

False portrayal of an organization, product or project as environmentally friendly to improve its public image or gain market advantages. In the context of carbon credits, greenwashing occurs when projects or companies exaggerate their environmental impact without delivering meaningful emissions reductions.

Nationally determined contributions (NDCs)

Country-specific climate action plans, under the Paris Agreement, that outline how countries plan to reduce their greenhouse gas emissions and adapt to climate change's impacts.

Permanence

Used to describe the longevity of emissions reductions or carbon sequestration. A carbon credit project should ensure that its emissions reduction or carbon sequestration is permanent, meaning future actions, such as deforestation or industrial activity, will not reverse it.

Proof of stake (PoS)

A blockchain consensus mechanism in which validators are chosen based on the amount of cryptocurrency they hold and their stake. Since it does not require mining, PoS is more energy-efficient than proof of work. It offers faster transaction processing with lower energy costs.

REDD+ (reducing emissions from deforestation and forest degradation)

A framework developed by the UN to combat climate change by incentivizing forest conservation and sustainable management. It focuses on reducing greenhouse gas emissions by preventing deforestation and forest degradation while promoting the sustainable use of forests and enhancing forest carbon stocks, especially in developing countries. It also emphasizes social and environmental safeguards to ensure benefits to local communities and biodiversity.

Reforestation
The process of replanting trees in areas where forests have been cleared or degraded. It helps to absorb CO_2 from the atmosphere, contributing to carbon sequestration efforts.

Renewable energy certificates (RECs)
Tradable certificates that represent proof that one mega-watt-hour of electricity was generated from a renewable energy source like wind or solar. Companies use RECs to offset their carbon footprint.

Standards governing bodies
Organizations that set the criteria and procedures for issuing, monitoring and verifying carbon credits. Examples include the Gold Standard, the Verified Carbon Standard and Verra. These bodies ensure the credibility of carbon credits but have been criticized for being slow to adapt to technological advancements.

Sustainable Development Goals (SDGs)
A set of 17 global goals established by the UN in 2015 to address urgent social, economic and environmental challenges, with a deadline in 2030. The SDGs include targets for climate action, poverty reduction and responsible consumption.

Traceability
The ability to track a carbon credit from its source through its entire lifecycle, ensuring transparency and accountability. Blockchain technology is often used to enhance traceability in the carbon credit market.

Voluntary carbon market
A market where companies and individuals voluntarily purchase carbon credits to offset their emissions, as opposed to compliance markets, where carbon credits are purchased to meet legal obligations.

ACKNOWLEDGEMENTS

As I come to the end of writing this book, I am deeply grateful for the incredible individuals who have shaped this journey and supported me in large and small ways. Without each of you, none of this would have been possible.

To my parents, Lord Thomas Stephen and Lady Margaret Emma Brasko – yes, they own one square foot of land in Scotland, which gave them titles – thank you for raising me to be strong, determined and committed to doing the right thing despite obstacles. You instilled the values and strength that have carried me through every challenge.

To my fantastic daughter, Kiley, your ever-loving support means the world to me. Your belief and constant encouragement have strengthened and inspired me. Travelling with you throughout the world has been one of the greatest gifts of my life. Thank you for being by my side every step of the way.

Holly Smithson and Britt Styr, my best friends and biggest supporters – you have always been there to cheer, celebrate, laugh and cry with me. Your friendship has been a lifeline, and I could not have asked for better companions on this journey.

Stanley Mathuram, your insights and relentless pursuit of real impact in this field have been a guiding light. You continue to inspire me with your leadership and commitment to pushing bold boundaries.

Steve and Wendy Edwards and Damien Mander, unsung heroes in conservation, your tireless work on the front lines deserves recognition. Your efforts give real meaning to the fight against corruption in the name of climate change.

Chief Mola and the incredible communities that have trusted me to share their stories – thank you for your courage, resilience and unwavering commitment to the truth. This book is dedicated to you and the Mola village's future.

Dr William Dewar passed away in December 2024. I am thankful he shared his profound knowledge of climate science and physical oceanography. His expertise was invaluable in shaping my understanding of carbon capture. Our conversation helped to unlock for me the potential of natural solutions like hemp and biochar. I also thank him for putting his son Vince into my life to provide travel adventures filled with humour throughout Africa.

Tom Miles, for your dedication to advancing biochar technologies, and Karl Strahl at Oregon Biochar Solutions, thank you for showing me the practical applications of biochar in carbon sequestration. Your work is helping to drive real, measurable progress.

Jamie Bartley of Unyte, your commitment to integrating sustainable solutions is inspirational. Thank you for helping me understand how to scale regenerative practices to create long-term impact.

Tafadzwa Dutoit Nyamande, your incredible dedication and pioneering spirit in biochar carbon removal have left a lasting impression. Your unwavering commitment to transforming agricultural waste into sustainable solutions and innovative vision for Africa's circular economy are truly inspiring. Thank you for welcoming me into your world, sharing your invaluable knowledge, and showing me how determination, ingenuity and community can make a global impact. Here's to many more groundbreaking adventures!

Anete Garoza, you have been an incredible partner in advocating transparency and integrity in nature-based carbon projects. Your work with 1MT Nation is transforming lives and restoring ecosystems. I am honoured to work alongside you in the fight for climate justice.

Daphne de Jong, your insights into renewable energy and your unstoppable determination continue to inspire me. From Mount Everest to pioneering technology, your resilience is unmatched. I am looking forward to Greenland and Mongolia. Thank you for helping me understand the intricacies of energy markets and for showing me how we can push boundaries in both technology and impact.

I want to express my heartfelt thanks to the teams at Fiùtur, RippleNami and Caelum Resources. Your innovation and drive have enabled us to reimagine the carbon credit market and create real, measurable change.

Audrey Jacobs, thank you for coming up with the unforgettable TEDx San Diego opportunity and the title "Carbon Credits Are Crap." Your creativity sparked the perfect way to bring this message to life.

To the organizers of the Africa Voluntary Carbon Credits Market Forum event in Zimbabwe: you still owe me and my partners a lot of money for our expenses. Yet, this is another reminder of how the carbon credit market needs reform – on all fronts.

I want to extend my heartfelt thanks to Sally Percy and Daniel Shaffer for your encouragement and the generous introduction to LID Publishing. Your support truly means a lot. And a big thank you to Martin Liu and the entire team at LID for your belief in my work and the opportunity to bring these ideas to life. I'm deeply grateful.

And to everyone else who has been part of this journey – whether your contributions were direct or through your support and encouragement – I am grateful beyond words. We are shaping a more transparent, equitable and impactful future. Thank you for being a part of this mission to turn those crappy carbon credits into gold for the people who need it most.

FURTHER READING AND RESOURCES

For readers interested in exploring the topics explored in this book, here are a few recommended resources, books and organizations that offer valuable insights into carbon credits, climate change, sustainability and global development.

BOOKS AND REPORTS

Standardising Carbon Markets by Springer Nature (2025)
A comprehensive guide (to which I am a contributing author) to the evolving landscape of carbon markets, detailing standards, methodologies, and best practices for achieving transparency and accountability in carbon trading.

The Uninhabitable Earth: Life After Warming by David Wallace-Wells (2019)
A powerful and sobering exploration of the potential impacts of climate change and the urgent need for collective action.

Climate Justice: Hope, Resilience, and the Fight for a Sustainable Future by Mary Robinson (2018)

Former UN High Commissioner for Human Rights Mary Robinson shares stories of people who have faced and responded to the challenges of climate change with resilience and hope.

This Changes Everything: Capitalism vs. The Climate by Naomi Klein (2014)

A compelling examination of the relationship between free-market capitalism and the climate crisis, arguing for systemic change to tackle global warming.

Green Swans: The Coming Boom in Regenerative Capitalism by John Elkington (2020)

This book explores how regenerative capitalism could be the key to tackling climate change, emphasizing sustainability and long-term economic resilience.

State and Trends of Carbon Pricing by the World Bank

A yearly report that provides comprehensive updates on global carbon pricing, including carbon markets and taxes. Visit www.worldbank.org.

ORGANIZATIONS

Fiùtur

A technology-driven company focused on revolutionizing the carbon credit market through innovative monitoring, verification and traceability solutions.
Visit www.fiuturx.com.

RippleNami

A global data and technology company focused on delivering transparent, traceable solutions in carbon credit markets and sustainable development for governments and organizations.
Visit www.ripplenami.com.

1MT Nation

An organization focused on removing 1 million metric tonnes of carbon from the atmosphere through high-quality, nature-based carbon removal projects in Africa.
Visit www.1mtn.com.

World Resources Institute

A global research organization that works on environmental sustainability, including climate change, carbon markets and natural resource management.
Visit www.wri.org.

United Nations Framework Convention on Climate Change (UNFCCC)

The UN is responsible for global climate negotiations and frameworks such as the Paris Agreement.
Visit www.unfccc.int.

WEBSITES AND ARTICLES

Carbon Brief
A UK-based website providing clear, in-depth information about the latest climate science, policy and carbon market developments.
Visit www.carbonbrief.org.

Project Drawdown
A comprehensive resource on the most effective climate solutions, covering areas from renewable energy to carbon sequestration through sustainable agriculture.
Visit www.drawdown.org.

Science Based Targets Initiative
An initiative that helps companies set ambitious climate targets aligned with the latest science, focusing on decarbonization and sustainability.
Visit www.sciencebasedtargets.org.

TEDx Talks on Climate Change and Carbon Credits
Access a wealth of TEDx talks that explore the intersections of carbon credits, climate action and global equity, including my talk "Carbon Credits Are Crap."
Visit www.ted.com/tedx.

PODCASTS

Outrage + Optimism
This podcast, hosted by Christiana Figueres, Tom Rivett-Carnac and Paul Dickinson, features discussions on climate change, politics and the challenges of reducing global emissions.
Visit www.outrageandoptimism.org.

The Energy Transition Show
This podcast focuses on the transition to renewable energy and how it intersects with carbon markets, economic policy and technological innovation.
Visit https://xenetwork.org/ets.

Carbon Copy
A podcast that tracks the biggest news stories in climate change and how they affect business, technology and markets, including carbon credits.
Visit https://carboncopy.eco/podcast.

ENDNOTES

1. Jaye Connolly, "Carbon Credits Are Crap: The Hidden Impact on African Communities," (TEDx Talks), last modified 19 January 2025, https://www.youtube.com/watch?v=0VTQ_RtK3y8.

2. *Carbon Credit Market: Analysis by Traded Value, Traded Volume, Segment, Project Category, Region, Size and Trends with Impact of COVID-19 and Forecast up to 2028*, (Research and Markets, 2023), accessed 26 January 2025, https://www.researchandmarkets.com/reports/5774731/carbon-credit-market-analysis-traded-value?srsltid=AfmBOorc-gLLiVLpTRZMOd6cznvORq7OIPXMG_JkaWgcnOHY9cJXGtWm.

3. Nick Routley, "Mapped: Visualizing the True Size of Africa," (*Visual Capitalist*), last modified 6 February 2024, https://www.visualcapitalist.com/mapped-africas-population-density-patterns.

4. Bart Crezee and Ties Gijzel, "Showcase Project by the World's Biggest Carbon Trader Actually Resulted in More Carbon Emissions," (Follow the Money), last modified 27 January 2023, https://www.ftm.eu/articles/south-pole-kariba-carbon-emission.

5. See "Gonarezhou National Park," (Africaproof), accessed 26 January 2025, https://africaproof.com/gonarezhou-national-park.

6. See https://musangosafaricamp.com.

7. James Ashworth, "Fossil Hunter Discovers New Species of 210-Million-Year-Old Lungfish," (Natural History Museum), last modified 6 August 2024, https://www.nhm.ac.uk/discover/news/2024/august/fossil-hunter-discovers-new-species-210-million-year-old-lungfish.html.

8. See https://www.akashinga.org.

9. See https://films.nationalgeographic.com/akashinga.

10. "Diverse Languages of Africa," (Africa View Facts), last modified 2 May 2024. https://africaviewfacts.com/article/diverse-languages-of-africa.

11. "Income Inequality in Africa: Literature Review and Data Gaps," (African Development Bank Group), last modified 16 February 2025, https://www.afdb.org/fileadmin/uploads/afdb/Documents/Generic-Documents/Revised-Income%20inequality%20in%20Africa_LTS-rev.pdf.

12. "Facts and Figures: Economic Development in Africa Report 2021," (United Nations Conference on Trade and Development), last modified 31 August 2024, https://unctad.org/press-material/facts-and-figures-7.

13. Lee J. T. White, Eve Bazaiba Masudi, Jules Doret Ndongo, Rosalie Matondo, Arlette Soudan-Nonault, Alfred Ngomanda, Ifo Suspense Averti, Corneille E. N. Ewango, Bonaventure Sonké and Simon L. Lewis, "Congo Basin Rainforest: Invest US$150 Million in Science," (*Nature*), last modified 20 October 2021, https://www.nature.com/articles/d41586-021-02818-7

14. Hervé Demarcq and Laila Somoue, "Phytoplankton and Primary Productivity off Northwest Africa," in: *Oceanographic and Biological Features in the Canary Current Large Marine Ecosystem*, eds. Luis Valdes and Itahisa Déniz-González (UNESCO, 2015), accessed 26 January 2025, https://unesdoc.unesco.org/ark:/48223/pf0000258919.

15. Thomas Pakenham, *The Scramble for Africa* (New York: Random House, 1991).

16. Adam Hochschild, *King Leopold's Ghost: A Story of Greed, Terror, and Heroism in Colonial Africa* (Boston: Houghton Mifflin Harcourt, 1998).

17. "IMF Austerity Loan Conditions Risk Undermining Rights," (Human Rights Watch), last modified 25 September 2023, https://www.hrw.org/news/2023/09/25/imf-austerity-loan-conditions-risk-undermining-rights.

18. *Climate Finance Unchecked: How Much Does the World Bank Know about the Climate Actions It Claims?* (Oxfam, 2024), last modified 26 January 2025, https://oxfamilibrary.openrepository.com/bitstream/handle/10546/621658/bp-climate-finance-unchecked-241017-en.pdf.

19. *"A New Era in Development: Annual Report 2023,"* (World Bank, 2023), last modified 26 January 2025, https://documents1.worldbank.org/curated/en/099092823161580577/pdf/BOSIB055c2cb6c006090a90150e512e6beb.pdf.

20. Jane Flanagan, "The Kenyan Train to Nowhere Reveals China's Debt-Trap Diplomacy," (*The Times*), last modified 13 August 2024, https://www.thetimes.com/world/africa/article/kenyan-train-to-nowhere-reveals-chinas-debt-trap-diplomacy-h2j0qrtqs.

21. Sha Hua and Gabriele Steinhauser, "China Shores Up Ties with Africa Despite Slowing Economy and Friction Over Debt," (*The Wall Street Journal*), last modified 5 September 2024, https://www.wsj.com/world/china-shores-up-ties-with-africa-despite-slowing-economy-and-friction-over-debt-94079613.

22. Philipp Sander, "Do Chinese Firms Employ Convicts from China in Africa?" (*Deutsche Welle*), last modified 22 December 2023, https://www.dw.com/en/do-chinese-firms-employ-convicts-from-china-in-africa/a-67802241.

23. Célia Cuordifede, "Au Sénégal, des ouvriers face au revers de la médaille économique chinoise," (*Le Monde*), last modified 6 September 2024, https://www.lemonde.fr/afrique/article/2024/09/05/au-senegal-des-ouvriers-face-au-revers-de-la-medaille-economique-chinoise_6305013_3212.html.

24. Ivy S. So, Barbara K. Haya and Micah Elias, *Voluntary Registry Offsets Database v8,* (Berkeley Carbon Trading Project), last modified 10 May 2023, https://gspp.berkeley.edu/research-and-impact/centers/cepp/projects/berkeley-carbon-trading-project/offsets-database.

25. Ivy S. So, Barbara K. Haya and Micah Elias, *Voluntary Registry Offsets Database v8*, (Berkeley Carbon Trading Project), last modified 10 May 2023, https://gspp.berkeley.edu/research-and-impact/centers/cepp/projects/berkeley-carbon-trading-project/offsets-database.

26. Bart Crezee and Ties Gijzel, "Showcase Project by the World's Biggest Carbon Trader Actually Resulted in More Carbon Emissions," (Follow the Money), last modified 27 January 2023, https://www.ftm.eu/articles/south-pole-kariba-carbon-emission.

27. "Where the Carbon Offset Market Is Poised to Surge," (Morgan Stanley), last modified 11 April 2023, https://www.morganstanley.com/ideas/carbon-offset-market-growth.

28. "EU Emissions Trading System (EU ETS)," (European Commission), accessed 26 January 2025, https://climate.ec.europa.eu/eu-action/eu-emissions-trading-system-eu-ets_en.

29. "Where the Carbon Offset Market Is Poised to Surge," (Morgan Stanley), last modified 11 April 2023, https://www.morganstanley.com/ideas/carbon-offset-market-growth.

30. Anders Porsborg-Smith, Jesper Nielsen, Bayo Owolabi and Carl Clayton, "The Voluntary Carbon Market is Thriving," (Boston Consulting Group, 2023), accessed 24 August 2024, https://www.bcg.com/publications/2023/why-the-voluntary-carbon-market-is-thriving.

31. "The Clean Development Mechanism," (United Nations Climate Change), accessed 25 January 2025, https://unfccc.int/process-and-meetings/the-kyoto-protocol/mechanisms-under-the-kyoto-protocol/the-clean-development-mechanism.

32. "Validation and Verification Bodies," (Gold Standard), accessed 26 January 2025, https://globalgoals.goldstandard.org/verification-validation-bodies.

33. "Verified Carbon Standard," (Verra), accessed 26 January 2025, https://verra.org/programs/verified-carbon-standard.

34. Regan, Helen, "Most Carbon Credits Are 'Phantom Credits' and Bad for the Environment, Study Shows," (TIME), last modified 27 March 2023. https://time.com/6264772/study-most-carbon-credits-are-bogus/.

35. Ben Elgin, Alastair Marsh and Max de Haldevang, "Faulty Credits Tarnish Billion-Dollar Carbon Offset Seller," (Bloomberg), last modified 24 March 2023, https://www.bloomberg.com/news/features/2023-03-24/carbon-offset-seller-s-forest-protection-projects-questioned.

36. Bart Crezee and Ties Gijzel, "Showcase Project by the World's Biggest Carbon Trader Actually Resulted in More Carbon Emissions," (Follow the Money), last modified 27 January 2023, https://www.ftm.eu/articles/south-pole-kariba-carbon-emission.

37. Patrick Greenfield, "Revealed: More than 90% of Rainforest Carbon Offsets by Biggest Provider Are Worthless, Analysis Shows," (*The Guardian*), last modified 18 January 2023, https://www.theguardian.com/environment/2023/jan/18/revealed-forest-carbon-offsets-biggest-provider-worthless-verra-aoe.

38. Heidi Blake, "The Great Cash-for-Carbon Hustle," (*The New Yorker*), last modified 16 October 2023, https://www.newyorker.com/magazine/2023/10/23/the-great-cash-for-carbon-hustle.

39. "The Carbon Con," (SourceMaterial), last modified 18 January 2023, https://www.source-material.org/vercompanies-carbon-offsetting-claims-inflated-methodologies-flawed.

40. Simon Counsell, *Blood Carbon: How a Carbon Offset Scheme Makes Millions from Indigenous Land in Northern Kenya* (Survival International, 2023), accessed 26 January 2025, https://assets.survivalinternational.org/documents/2466/Blood_Carbon_Report.pdf.

41. Ben Elgin, Alastair Marsh and Max de Haldevang, "Faulty Credits Tarnish Billion-Dollar Carbon Offset Seller," (Bloomberg), last modified 24 March 2023, https://www.bloomberg.com/news/features/2023-03-24/carbon-offset-seller-s-forest-protection-projects-questioned.

42. Simon Counsell, *Blood Carbon: How a Carbon Offset Scheme Makes Millions from Indigenous Land in Northern Kenya* (Survival International, 2023), accessed 26 January 2025, https://assets.survivalinternational.org/documents/2466/Blood_Carbon_Report.pdf.

43. Heidi Blake, "The Great Cash-for-Carbon Hustle," (*The New Yorker*), last modified 16 October 2023, https://www.newyorker.com/magazine/2023/10/23/the-great-cash-for-carbon-hustle.

44. Bart Crezee and Ties Gijzel, "Showcase Project by the World's Biggest Carbon Trader Actually Resulted in More Carbon Emissions," (Follow the Money), last modified 27 January 2023, https://www.ftm.eu/articles/south-pole-kariba-carbon-emission.

45. See https://carbongreenafrica.net.

46. See https://carbongreeninvestments.com.

47. See https://www.southpole.com.

48. "Who We Are," (Verra), accessed 26 January 2025, https://verra.org/about/overview.

49. "Carbon Offset Verification," (SCS Global Services), accessed 26 January 2025, https://www.scsglobalservices.com/services/carbon-offset-verification.

50. "South Pole Ends Agreements with Carbon Green Investments (CGI), Owner of Kariba REDD+ Project," (South Pole), last modified 27 October 2023, https://www.southpole.com/news/statement-27october.

51. See https://www.northernkenyacommunitycarbon.org.

52. See https://www.nrt-kenya.org.

53. "USAID Partnership with Northern Rangelands," (US Agency for International Development), accessed 26 January 2025, https://www.usaid.gov/kenya/document/usaid-partnership-northern-rangelands.

54. "Update: Northern Kenya Grassland Carbon Project," (Verra), last modified 26 March 2023, https://verra.org/program-notice/update-northern-kenya-grassland-carbon-project.

55. Simon Counsell, *Blood Carbon: How a Carbon Offset Scheme Makes Millions from Indigenous Land in Northern Kenya*, (Survival International, 2023), accessed 26 January 2025, https://assets.survivalinternational.org/documents/2466/Blood_Carbon_Report.pdf.

56. "Rejoinder to NRT's Statement on the *Blood Carbon: How a Carbon Offset Scheme Makes Millions from Indigenous Land in Northern Kenya* Report," (Borana Council of Elders), accessed 26 January 2025, https://assets.survivalinternational.org/documents/2476/230327_Borana_Council_of_Elders_R_letter.pdf.

57. "NRT Is Doing the Wrong Thing that I Find Unjust to Our People," (Survival International), last modified 16 March 2023, https://www.youtube.com/watch?v=2iNxMz6VAu8.

58. "United States Seed Funding Yields Community Impact for Conservation, People and Landscapes," (US Agency for International Development, 2022), accessed 26 January 2025, https://www.usaid.gov/sites/default/files/2022-05/NRT_Factsheet_CORRECTED_April_2022.pdf.

59. "The Voluntary Carbon Market in 2024," (Abatable), last modified 12 February 2025. https://abatable.com/blog/the-voluntary-carbon-market-in-2024/.

60. Kenza Bryan, "Carbon Credits from Cookstove Emissions Largely Worthless, Study Finds," (*Financial Times*), last modified 23 January 2024, https://www.ft.com/content/6a9d7ef7-2e30-4082-8ae0-3a722008ddab.

61. Annelise Gill-Wiehl, Daniel M. Kammen and Barbara K. Haya, "Pervasive Overcrediting from Cookstove Offset Methodologies," *Nature Sustainability* 7 (2024): 191–202. https://rael.berkeley.edu/wp-content/uploads/2024/07/Gill-Wiehl-Kammen-Haya-Cooking-the-Books-Nature-Sustiainability-2024.pdf.

62. "Open Letter: Experts Challenge Misguided Criticism of Clean Cookstoves Funding," (C-Quest Capital, 2023), accessed 26 January 2025, https://cquestcapital.com/wp-content/uploads/2023/10/Cookstoves-open-letter.pdf.

63. Jennifer L, "Former C-Quest Capital CEO Accused of $100m Carbon Credit Fraud Scheme," (Carbon Credits), last modified 7 October 2024, https://carboncredits.com/former-c-quest-capital-ceo-accused-of-100m-carbon-credit-fraud-scheme.

64. Judith Simon, "On Today's Release from C-Quest Capital Regarding Allegations Against Former C-Quest CEO Ken Newcombe," (Verra), last modified 26 June 2024, https://verra.org/c-quest-capital-statement.

65. "Germany Blocks CO2 Vouchers for Oil Companies over Fraud Concerns in China," (Reuters), last modified 6 September 2024, https://www.reuters.com/business/energy/germany-blocks-co2-vouchers-oil-companies-over-fraud-concerns-china-2024-09-06.

66. "Case Study: How Can dMRV Support High Quality Carbon Credits?" (Senken), last modified 21 August 2024, https://www.senken.io/blog/case-study-how-can-dmrv-support-high-quality-carbon-credits.

67. "Measuring Forest Carbon the Old-School Way," (PBS Terra) YouTube video, last modified 12 October 2023. https://www.youtube.com/watch?v=AGPkF3T2NNE.

68. See "Our Mission" (Fiùtur), accessed 26 January 2025, https://www.fiuturx.com/our-vision.

69. Julian Milnes, "HFC Reclaim Could Save 18 Bn Tonnes of CO_2," (RAC Plus), last modified 29 April 2015. https://www.racplus.com/news/hfc-reclaim-could-save-18-bn-tonnes-of-co2-29-04-2015.

70. "Solution" (Fiùtur), accessed 26 January 2025, https://www.fiuturx.com/solution.

71. "Who We Are," (RippleNami), accessed 26 January 2025, https://ripplenami.com/who-we-are.

72. "RippleNami Carbon Clarity™ System," (RippleNami), accessed 26 January 2025, https://ripplenami.com/carbon-clarity.

73. "Farm Bill," (US Department of Agriculture), accessed 26 January 2025, https://www.usda.gov/farming-and-ranching/farm-bill.

74. Darren Kaplan, "The Evolution of Cannabis: Why Was Hemp Made Illegal?" (Clark Hill), last modified 28 August 2020, https://www.clarkhill.com/news-events/news/the-evolution-of-cannabis-why-was-hemp-made-illegal.

75. Andrew Leonard, "Can Hemp Clean Up the Earth?" (Rolling Stone), last modified 11 June 2018, https://www.rollingstone.com/politics/politics-features/can-hemp-clean-up-the-earth-629589.

76. "The OBS Difference," (Oregon Biochar Solutions), accessed 26 January 2025, https://www.chardirect.com/about-oregon-biochar-solutions.

77. See https://unyte.co.uk.

78. See https://www.zimbanjex.com.

79. See https://www.caelumresources.com.

80. See https://avccmforum.org.

81. "Latvia – Forest Area (% of Land Area)," (Trading Economics), last accessed 19 February 2025, https://tradingeconomics.com/latvia/forest-area-percent-of-land-area-wb-data.html.

82. "Latvia: Regional Analysis," (Carbon Gap Tracker), last accessed 19 February 2025, https://tracker.carbongap.org/regional-analysis/national/latvia.

83. See https://www.1mtn.com.

84. "Bamboo Produces More Oxygen Than Trees," (Aussie Bamboo), accessed 19 February 2025, https://www.aussiebamboo.com.au/bamboo-produce-more-oxygen-than-trees.

85. Erick Salas Burgueño, "Global Forest Size 1990-2021," (Statista), accessed 20 August 2024, https://www.statista.com/statistics/1292175/global-forest-area/#:~:text=Global%20forest%20size%201990%2D2021&text=Forests%20cover%20over%20four%20billion,percent%20of%20total%20land%20area.

86. "Mozambique Becomes First Country to Receive Emission Reductions Payments from Forest Carbon Partnership Facility," (World Bank), last modified 15 October 2021, https://www.worldbank.org/en/news/press-release/2021/10/15/mozambique-becomes-first-country-to-receive-emission-reductions-payments-from-forest-carbon-partnership-facility.

87. "ZILMP – Background Study for the Preparation of the Zambezia Integrated Landscapes Management Program," (Nitidæ), accessed 19 February 2025, https://www.nitidae.org/en/actions/zilmp-etude-de-preparation-a-un-programme-juridictionnel-redd-dans-la-province-de-zambeze-zambezia-integrated-landscapes-management-program.pdf.

88. "Commodities at a Glance: Special Issue on Access to Energy in Sub-Saharan Africa," (United Nations Conference on Trade and Development), access date 19 February 2025, https://unctad.org/publication/commodities-glance-special-issue-access-energy-sub-saharan-africa#:~:text=Globally%2C%20733%20million%20people%2C%20or,live%20without%20access%20to%20electricity.

89. See https://www.vencover.com.

90. 'Profile: Daphne de Jong,' (Forbes), accessed 26 January 2025, https://www.forbes.com/profile/daphne-de-jong.

91. "Africa Cannot Confront Climate Change Alone," (International Monetary Fund), accessed 17 December, 2024, https://www.imf.org/en/Blogs/Articles/2021/12/17/africa-cannot-confront-climate-change-alone.

92. "COP29 Unpacked: Wins, Losses, and Controversies," (European University Institute), accessed 19 December 2024, https://www.eui.eu/news-hub?id=cop29-unpacked-wins-losses-and-controversies.

93. "The Clean Development Mechanism," (United Nations Climate Change), accessed 26 January 2025, https://unfccc.int/process-and-meetings/the-kyoto-protocol/mechanisms-under-the-kyoto-protocol/the-clean-development-mechanism.

94. "EU Emissions Trading System (EU ETS)," (European Commission), accessed 26 January 2025, https://climate.ec.europa.eu/eu-action/eu-emissions-trading-system-eu-ets_en.

95. "Nationally Determined Contributions (NDCs)," (United Nations Climate Change), accessed 26 January 2025, https://unfccc.int/process-and-meetings/the-paris-agreement/nationally-determined-contributions-ndcs.

96. Matt Hussey, "Crypto's Carbon Emissions Problem and the Projects Trying To Solve It," (Decrypt), accessed 19 February 2025, https://decrypt.co/63621/cryptos-carbon-emissions-problem-and-the-projects-trying-to-solve-it.